A Sense of Place Book 2

A Sense of Place

Book 2 Places, Resources and People

Rex Beddis

Oxford University Press

Oxford University Press, Walton Street, Oxford OX2 6DP
Oxford   New York   Toronto
Delhi Bombay Calcutta Madras Karachi
Petaling Jaya Singapore Hong Kong Tokyo
Nairobi Dar es Salaam Cape Town
Melbourne Auckland

and associated companies in
Berlin Ibadan

Oxford is a trade mark of Oxford University Press

© Oxford University Press 1982

First published 1982
Reprinted 1982, 1983 (twice), 1985, 1986, 1987, 1988, 1989, 1990, 1991

All rights reserved. No part of this publication may be reproduced, stored in a retrieval system, or transmitted, in any form or by any means, without the prior permission in writing of Oxford University Press. Within the UK, exceptions are allowed in respect of any fair dealing for the purpose of research or private study, or criticism or review, as permitted under the Copyright, Designs and Patents Act, 1988, or in the case of reprographic reproduction in accordance with the terms of licences issued by the Copyright Licensing Agency. Enquiries concerning reproduction outside those terms and in other countries should be sent to the Rights Department, Oxford University Press, at the address above.

ISBN 0 19 833434 6

Typeset by Tradespools Limited, Frome, Somerset
Printed in Hong Kong.

Preface

This series brings together the key elements of physical, human, and environmental geography through the study of carefully selected regional and national examples. Thus, contexts for study are provided which range from the United Kingdom to Europe (with particular reference to the EC) and a selection of countries from the rest of the world including a balance of large and small and more and less developed countries. Through the study of themes set in locational frameworks a variety of geographical skills can be acquired, including elements of graphicacy. This series of books therefore represents a coherent course for pupils studying geography in Key Stage 3.

The approach of each book is to present a number of ideas through a range of data – text, photographs, drawings, diagrams, statistics, maps and so on. The emphasis is on the understanding of ideas rather than the memorisation of facts, although the acquisition of knowledge of places and peoples is regarded as an important aim. At the same time a wide range of skills should be developed while the themes encourage a clarification of attitudes and values. Pupils are encouraged not only to make judgements from evidence, but also to express their feelings and values however acquired.

Book 1 focuses on basic ideas about location, pattern and process with illustrations from Britain and the experience of British people. Book 2 is concerned with the relationships between environments, resources and people at scales ranging from local to global. In Book 3 the emphasis changes again to economic, social and political aspects of 'development', and to the nature and behaviour of nations and groups. In all three books the aim is to help pupils look at the future as well as understand the present.

Contents

1 **Planet earth**　　Earth in space　8
　　　　　　　　　　The earth's crust　10
　　　　　　　　　　Energy from the sun　12
　　　　　　　　　　The web of life　14

Places

2 **Rainforest**　　Amazonia: life in the jungle　16
　　　　　　　　　The rainforest environment　18
　　　　　　　　　Development or destruction?　20
　　　　　　　　　World patterns　22

3 **Desert**　　　　The Empty Quarter　24
　　　　　　　　　The desert environment　26
　　　　　　　　　Change in Saudi Arabia　28
　　　　　　　　　World patterns　30

4 **Monsoon lands**　North Indian village　32
　　　　　　　　　　The burst of the monsoon　34
　　　　　　　　　　Flood, cyclone and drought:
　　　　　　　　　　　a year in India　36
　　　　　　　　　　Cities of the northern plain　38

5 **Polar lands**　　Eskimo whale hunt　40
　　　　　　　　　　The Arctic　42
　　　　　　　　　　Change in the Arctic　44
　　　　　　　　　　Antarctica　46

6 **Mid-latitudes**　Contrasts in California　48
　　　　　　　　　　Mountain, plateau and plain　50
　　　　　　　　　　The Great Plains　52
　　　　　　　　　　Ocean margin and continental
　　　　　　　　　　　interior　54

7 **Mountain**　　　Land of the Incas　56
　　　　　　　　　　Peruvian Andes　58
　　　　　　　　　　Building up and wearing down　60
　　　　　　　　　　Volcanoes　62

Resources

8	**Water resources**	Water: the basic resource 64 The River Nile 66 Using the River Nile 68 Using the rivers of northern Siberia 70
9	**Crop resources**	Soil, climate and crops 72 New land and controlled climates 74 Rice: more cropland, richer yields 76 Tea: a plantation crop 78

10	**Animal resources**	Friends, and food 80 Twentieth century herders 82 Commercial animal farming 84 Animals at risk: who cares? 86
11	**Resources from the sea**	Fish: harvest of the oceans 88 Sea life in the southern oceans 90 Hunting the whale 92 Changes in British fishing 94

12	**From the earth's crust**	North Sea oil 96 North Sea oil: gains and losses 98 Moving the world's oil 100 How long can our minerals last? 102
13	**Human skill**	Manufacturing things from resources 104 Making motor cars 106 Intermediate technology 108 Making and using microcomputers 110

14	**Places as resources**	Islands in the sun 112 Holidays in the Alps 114 East African game parks 116 Places with a past 118
15	**Places, resources and people**	Signs of strain 120 Unequal shares 122 Too many people? 124 Looking ahead 126
	Acknowledgements	128

1 Planet earth

Earth in space

The earth as seen from an Apollo spaceship, in orbit round the moon. The surface of the moon is in the foreground, the earth is about 380 000 km away

For thousands of years men and women knew little about the world, other than their immediate surroundings. They thought the earth was flat, and they could not understand the movements of the moon, sun and stars, nor the causes of day and night and the changing seasons.

Several centuries ago a number of brilliant astronomers and mathematicians worked out the truth. The earth is not flat, and although it is the home of the human race it is not the centre of the sun and the planets. It is almost a sphere, and it revolves around the sun. The shape of the earth was proved when men first sailed and then flew around it, while in recent decades space travellers have been able to get far enough away to see the earth as a whole.

The earth is one of nine planets orbiting the sun. Together with various moons, and other materials (like lumps of rock called 'asteroids'), the sun and its orbiting planets makes up what is called the solar system. The planets are very different in size, appearance, what they are made of, distance from the sun, speed and pattern of movement. Mercury, for example, rotates on its axis so slowly that it takes 176 of our days to get from one sunrise to the next! Venus rotates on its axis from east to west – the opposite way to all the other planets. Jupiter is eleven times the diameter of the Earth, and is more than twice as heavy as the other planets put together, yet it is mostly hydrogen and helium gas! But all these planets are held in their orbits by the sun's gravitational attraction, and are part of the same system or family. Our sun of course is a star and is just one of many millions of stars in the universe. Some of these can be seen on a cloudless night, but many, many more cannot be seen with the naked eye. The earth is a very, very tiny part of the universe.

The Solar System. The nine planets and their moons orbit the sun in ellipses of different size

Relative distances of the planets from the Sun

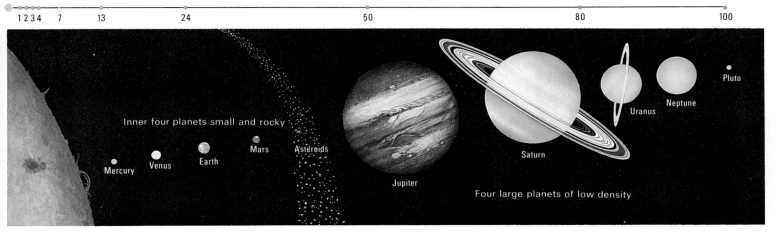

	Mean distance from the sun (million km)	Period of orbit	Period of rotation
Mercury	60	88.0 days	59 days
Venus	108	225.0 days	243 days
Earth	150	365.0 days	24 hours
Mars	228	687.0 days	24½ hours
Jupiter	778	11.9 years	10 hours
Saturn	1427	29.5 years	10¼ hours
Uranus	2870	84.0 years	11 hours
Neptune	4497	164.8 years	16 hours
Pluto	5900	247.7 years	6¼ days

The comparative sizes of the planets in relation to the sun

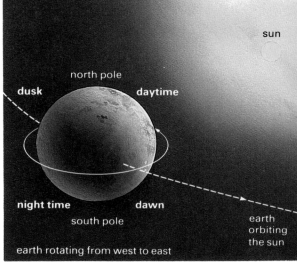

The earth rotates on its axis once in every twenty-four hours while orbiting the sun every 365¼ days

Although it is so small, our planet is extremely varied. This is partly due to the strange pattern of oceans and land masses called continents that we shall soon consider. It is also caused by the two main movements it makes – rotating on its axis every twenty four hours and orbiting the sun every 365¼ days.

1 Light travels at about 300 000 kilometres in one second. The average distance of the sun from the earth is about 150 000 000 kilometres (the earth orbits the sun in an ellipse, not a circle, and this distance is the average of the nearest and furthest distance). How long does it take light to travel from the sun to the earth?

2 Light travels about 10 000 000 000 000 (million million) kilometres in a year. The nearest fixed star (unlike 'shooting stars' or meteorites) is over 38 million million kilometres away. When we look at that star, how long has it taken the light we see to reach us?

3 Neither the earth nor the moon give off light – all our light comes from the sun. Yet we can often see a great deal at night by 'moonlight'. What actually is 'moonlight'?

4 Copy the diagram of the rotating earth. After the words 'dawn' and 'dusk' add the two following sentences, making sure they follow the most suitable of these two words.
Places rotate into the sun's rays – the sun appears to 'rise', bringing daylight.
Places rotate out of the sun's rays – the sun appears to 'set', bringing nightfall and darkness.

5 Imagine you were one of the astronauts seeing 'earthrise' over the edge of the moon. Describe a few of your thoughts or feelings.

The first American to walk in space, 3 June 1965

1 Planet earth   The earth's crust

This 'meteosat' (satellite) photograph of the earth shows the whole of Africa in clear detail as well as Spain, Portugal and the British Isles

The most striking feature of the earth when seen from a satellite or spacecraft is the contrast between vast land masses known as continents and even larger sea areas known as oceans. It is only recently that it has been possible for pictures to be taken from spacecraft and satellites so that human beings have been able to see the outlines of continents from photographs or with their own eyes. But the shapes of the continents and oceans are easy to recognise, for they have been shown for centuries on maps and globes.

This pattern of continent and ocean, however, gives an incomplete picture of the earth. The earth is over 12 700 kilometres from centre to surface. From the measurement of earthquake waves a little is known about the interior of the earth – the inner core and the outer mantle. Considerably more is known about the thin outer skin or crust, although actual drillings rarely exceed about ten kilometres.

The earth is covered by a number of more or less rigid 'plates' of solid and heavy rock that fit together like a huge jigsaw. These 'float' on a layer of partly molten rock. The plates are covered by layers of surface rock that may be as thin as five kilometres under the oceans and over thirty kilometres thick in the continents.

There are three main sorts of boundary between these plates. In one of these margins or boundaries there is a welling up of molten rock or magma from the interior of the earth. This usually takes place in the middle of the ocean. As the plates move slowly apart the new rock is added to the crust on the sea floor, rather like toothpaste being spread, along the lines of the mid-ocean ridges. Earthquakes often occur along these margins.

In other margins the edge of the ocean plate dips under the margin of the continental plate. This movement causes deep-seated earthquakes, while masses of molten rock find their way to the surface to produce volcanic eruptions.

A cross section of ocean and continental margins to show volcanic activity and how the ocean plate dips under the margin of the continental plate

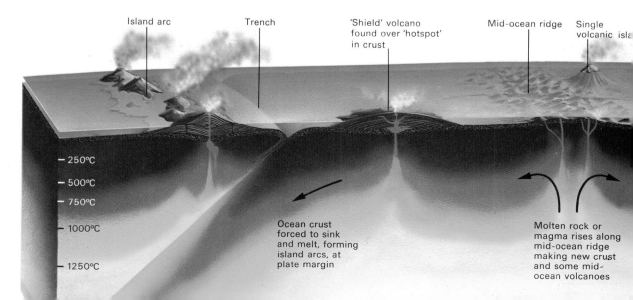

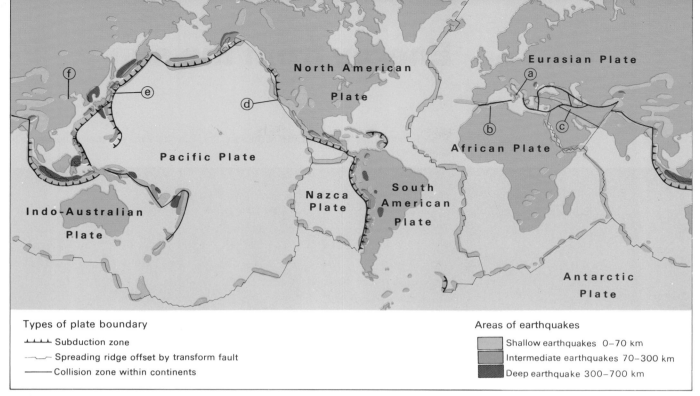

Types of plate boundary
- Subduction zone
- Spreading ridge offset by transform fault
- Collision zone within continents

Areas of earthquakes
- Shallow earthquakes 0–70 km
- Intermediate earthquakes 70–300 km
- Deep earthquake 300–700 km

Earthquakes can also occur in the third sort of movement where the edge of one plate moves horizontally and in the opposite direction to its neighbour. The best known example of this is the San Andreas fault system in California, the movement of which led to the disastrous San Francisco earthquake of 1912. The slow movement of plates, then, helps explain how new rock is added to the earth's crust, how mountains are formed and why the violent movements of earthquakes and volcanoes occur in some places rather than others.

1. Which continents and which oceans can be seen in the photograph of earth? Draw a picture based on the photograph and add in their names.
2. From the diagram below draw a profile to show the surface of the land, the ocean floor and the oceans. Use the map to label your profile: Pacific Plate, Nazca Plate, South American Plate, Andes fold mountains, New Zealand volcanoes.
3. There have been disastrous earthquakes at the places lettered a–f in recent years. Name the countries. Try and find out more about one of them – when it occurred and what were the results.

The earthquake zones and plate margins of the world. Intermediate and deep earthquakes are almost always found at where the ocean crust is forced to sink and under the edge of the continental plate, while shallow earthquakes also occur at other types of plate margin. Scattered quakes occur within plates in some locations

An aerial view of the effects of an earthquake in Alaska in 1964.

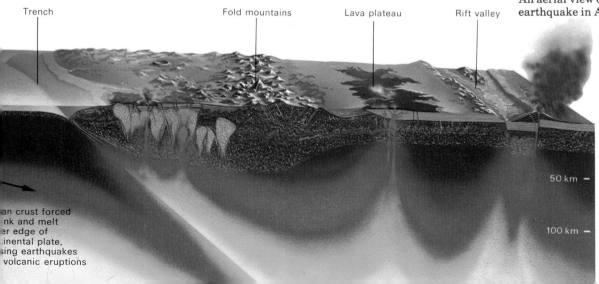

Trench — Fold mountains — Lava plateau — Rift valley

an crust forced
nk and melt
r edge of
inental plate,
sing earthquakes
volcanic eruptions

1   Planet earth   Energy from the sun

A 'heat photograph' of the surface of the sun. The solar flare may be many thousands of kilometres high

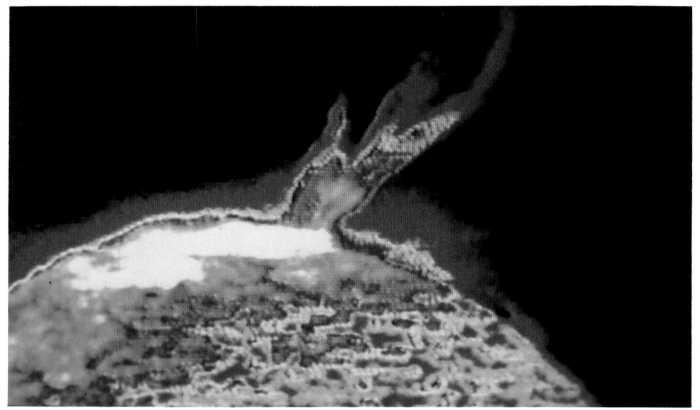

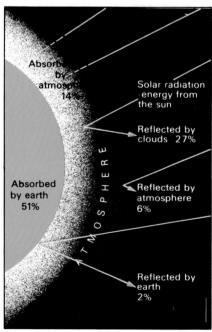

The earth's solar radiation budget

We have already seen that the earth receives all its light from the sun, either as direct sunlight or reflected moonlight. But sunlight is more important to us than just as a means of allowing us to see. Without sunlight plants would not grow, and without plant growth no life would be possible. The earth would be a dead planet.

Reactions taking place within the sun produce extremely high temperatures. From its glowing surface, heat and light energy is radiated out into space. The earth, at a distance of 150 000 000 kilometres from the sun, intercepts only a tiny fraction of this energy.

The earth is surrounded by a mixture of nitrogen, oxygen and many other important but small amounts of gases, known as the atmosphere. The atmosphere rapidly thins out away from the earth's surface. This is why mountaineers and people flying at any height need oxygen to breathe properly. All weather activities take place in the lower layers of the atmosphere, where there are often clouds.

The amount of energy received at the outer limits of the atmosphere is always about the same – it can vary a little with sunspot activity and with changing distance from the sun. But a lot of solar energy gets reflected back into space by the atmosphere and clouds. The remainder is absorbed and used by the atmosphere and plants and animals as we shall soon see. At night the earth radiates heat energy back out into space.

Because the atmosphere reflects and absorbs a lot of the solar energy, and because the earth's surface is curved, the amount of energy – heat and light – received at the surface varies a lot from place to place. The diagram shows how places where the sun is nearly overhead receive much more energy than those where the sun is usually low in the sky. It is because of this that places near the equator are usually hotter than places near the Poles. We shall see later on that it also explains why some months are hotter than others in most parts of the world – it is not usually because we are then nearer the sun!

1. Add up the energy reflected by the atmosphere and clouds and deflected by the earth. Next add up the energy absorbed by the atmosphere and the earth (used in heating the earth, creating weather and allowing plants to grow). Add the two totals together – is it the figure you expected?
2. Copy the diagram of energy falling on the earth's surface in two places. Beneath it write out the following, completing it in your own words. 'The same amount of solar energy will be received at 'a' and 'b' at the outer margin of the atmosphere. The amount of solar energy received at 'A' will be less than at 'B' partly because the area of surface to be covered is larger, and partly because . . . .'
3. Think of a day you have known when the sun had just risen or was about to set on a cloudless day. Now think of a day when the sun was high in a cloudless sky early on a summer afternoon. On which of these occasions was the sun hottest? Explain why.
4. Describe one way in which the energy from the sun can nowadays be 'captured' and used for heating.
5. Winds are the movement of air from one place to another. What provides the energy to make the air move? How is wind energy used by people?

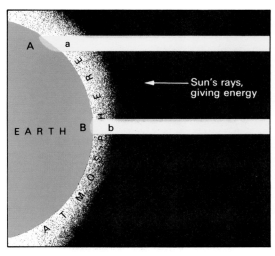

The effect of latitude on the energy received from the sun

Near the summit of Mt Everest. At these heights the direct sunlight is intense, but the thin air can be very cold

Left: A tornado whirls towards a small American town. Its track is narrow but buildings caught within it are smashed to the ground

1 Planet earth — The web of life

The huge River Amazon winds its way through dense tropical rainforest

All known life exists naturally within a very narrow zone close to the surface of the earth. There have been some very unusual occasions when men, other animals or plants have been shot into space for research purposes. These have been the only exceptions.

We have seen that the earth consists of large land masses and even larger oceans. Much of the land surface is covered with a thin layer of soil that teems with insect and animal life, and from which thousands of different sorts of plants get their food. Other plants are adapted to life in the sea, while great varieties of fish (and a few mammals) live in a water environment. Many animals wander over most parts of the earth's surface, though some species are adapted to different conditions, and some environments are too harsh to support any animal life. Birds have evolved to fly through the lower layers of the atmosphere, which cover the land and ocean surfaces. It is also in the lower levels of the atmosphere that great swirling masses of air move, creating our weather and climates. Each living thing depends on other living things, on the chemicals in the air, on the soil and water and on energy from the sun.

The way in which these different parts work together can be illustrated by the distribution and movement of water. Some water is held in the atmosphere as water vapour, cloud droplets or rain. Rather more is held in the soil and porous rocks near the surface. Some is held as ice in glaciers and ice sheets. A fairly small amount is found in lakes and rivers. By far the greatest amount is in the great oceans.

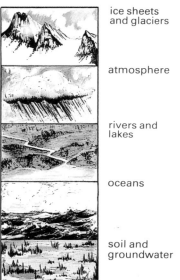

Where the earth's water is stored

The water cycle

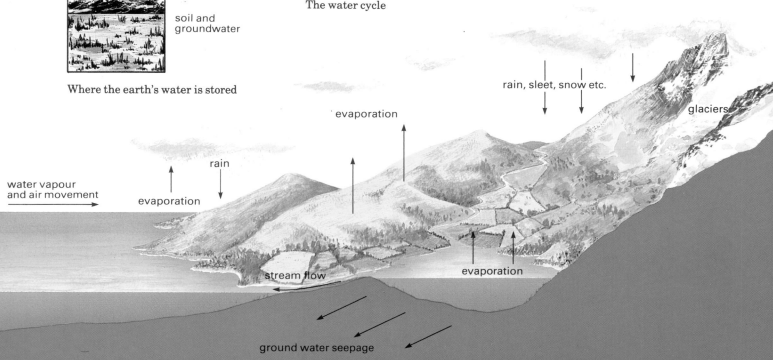

None of this water stays in one place, but is constantly moving around. Water evaporates from the surface of oceans, lakes, rivers ice sheets, glaciers and even moist soil. In the atmosphere the water vapour is moved around in masses of air until sooner or later the air is cooled enough for the vapour to turn back, or condense, into clouds, rain, sleet, hail or snow. This precipitation either falls on the oceans or the land. If it falls on the land it sinks into the ground to reappear again in some spring, river, lake or sea. Some of it will add to ice sheets and glaciers in cold climates. A lot will flow in channels as streams and rivers – and finally back into the oceans. This movement is known as the water cycle, and it is 'driven' by the sun's energy.

Water is needed by plants and animals, and there can be no life without water. Water also erodes and builds up the land, and moving water transports great masses of material around the earth. Later on we shall see how life depends on other cycles, but this one example should show how life on earth depends on a delicate balance of events.

1 Look at the diagram of 'Where the earth's water is stored' and at these approximate figures for the amounts involved.
   Atmosphere 0.008 per cent; oceans 97.2 per cent; ice sheets and glaciers 2.15 per cent; soil and ground water 0.625 per cent; rivers and lakes 0.017 per cent.
   Redraw the diagram putting the places in order, the largest percentage first.
2 In as few words as possible, describe how the 'water cycle' operates.
3 Read the extract carefully. How does the ozone layer affect life on Earth? What is causing damage to the ozone layer and how is it doing it? What might the effect be? What do you think should be done about it? Do you think it is true – and why?
4 Describe one small place you know, such as a hedge, pond or garden, to show how the 'parts' such as the weather, soil, plants, insects or fish 'fit' together to make an 'ecosystem'. Compare summer and winter conditions.

New fear as ozone layer dissolves

The National Academy of Sciences will issue a report this month saying that the ozone layer is breaking down at twice the rate scientists expected.

The report will say new calculations on fluorocarbon chemicals being discharged into the upper atmosphere suggest they are enough to break apart a little more than 14 per cent of the ozone layer in the next 50 to 100 years.

This is twice what the academy said in 1976 would be stripped from the ozone layer, which lies 28 miles above the earth and blocks most of the ultraviolet light.

Scientists are concerned that fluorocarbon gases used as propellants in spray cans accumulate in the upper atmosphere, where they are broken down by ultraviolet light and release chlorine that destroys the ozone. Scientists are worried that, if people continue to use spray cans containing fluorocarbon propellants as they did in 1975, about 7 per cent of the ozone layer would be stripped away in 50 years.

This would allow 14 per cent of the ultraviolet light from the sun to penetrate the atmosphere and cause a worldwide increase in skin cancer.

An extract from an article in the *Guardian* of 5 November 1979

What features in this scene from the African savanna help to illustrate the balance of nature?

Amazonia: life in the jungle

Above: An aeroplane casts a small shadow on the vast Amazonian rainforest

The forests in the photograph are part of a vast area more than ten times the size of England – Amazonia in South America. The Indians are from one of the few tribes that remain in these hot, humid jungles. No one quite knows where their ancestors came from, or when they first appeared in these huge forests.

Most of the things they used were made from forest materials, and these quickly decay, leaving little trace of the past. They also move from place to place in the forests, so remains of past villages are hard to find. It is thought that Indians have been living in this way in the Amazon forest for thousands of years.

The Indians adapted very cleverly to the jungle environment, and they managed to make use of every aspect of it without causing any major change. They obtained food by hunting animals, fishing in the many rivers, gathering wild produce and growing crops such as manioc, maize, yams, groundnuts and bananas. They grew these crops in patches of forest cleared by the 'slash-and-burn' method. (A modern large-scale and damaging version of this can be seen on the next page.) When there were only a few people this did no permanent damage to the forest. After a few years new 'gardens' would be made and the old ones revert to jungle. Many different sorts of fish and animals were caught, such as wild pigs, monkeys and tapir, with harpoons, spears, nets, blow-guns and dogs.

All the resources they used were from the surrounding forest. Wood was used for house frames, simple furniture and weapons, while clay was made into pottery. Their houses were thatched with palm

An Indian camp in a forest clearing. This will be their home for a short time. They will then move on to a new 'village'. These settlements are rapidly disappearing as the forest is destroyed

leaves, and hammocks, baskets and bow-strings were made from various sorts of reeds and grasses. Other foods such as fruits and honey were gathered from wild plants and trees. In one sense this was a very simple way of life, but in another it showed how people could cleverly adapt to a particular environment. Apart from hunting, farming and fishing, though, it seems there was a great deal of raiding and warfare between groups. This was not to take land, but to steal women, take prisoners and for head-hunting.

When Europeans first visited the Amazon rain forests several hundred years ago it is thought there were about two million forest Indians. Now there are less than 200,000. A large number died fighting the invaders, or being massacred by them. But far more died from diseases brought from Europe. The Indians had no resistance to diseases that were widespread but not fatal in Europe, and millions died from them. Before we see how the life of the Indians has changed, and what is happening to Amazonia, we will look more closely at why forests like this exist at all in this and similar parts of the world.

1. The map shows the size of the Amazon rainforest and the countries that include some of it. Which countries include some of Amazonia, and which of these has by far the largest area?
2. If you could travel by car at an average rate of, say, 500km a day – about the distance from London to Edinburgh – how many days would it take to cross Amazonia from Iquitos on the upper Amazon river to Belem at its mouth?
3. Imagine you were a member of an Indian village before it was affected by contact with Europeans. What would you have liked and what would you have disliked about the life? Do you think you would have had the skills to live in a community in such an environment?
4. What things do you and your family use that have *not* been obtained, changed or made by other people? Name a few things used by you and your family that have been made elsewhere.

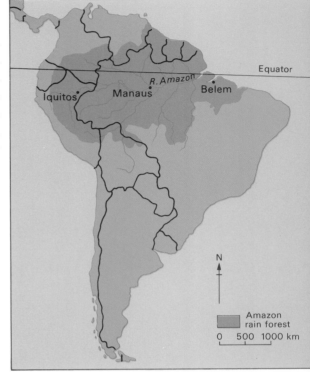

The rain forests of South America

A group of Xingu Indians with a friendly tapir

The rainforest environment

A tropical storm threatens over Bandung in Java, Indonesia

Logs placed on the swampy forest floor make a pathway for this young girl

The Amazon rainforest (or Amazonia) covers a huge lowland area drained by the River Amazon and its tributaries. This vast river rises in the Andes mountains to the west of Amazonia and, with its many tributaries, carries an enormous volume of water and load of silt into the Atlantic Ocean.

The area lies across the equator, which means that the mid-day sun is nearly overhead throughout the year. We have already discovered that a high sun results in a lot of energy being received by the surface of the earth and the lower layers of the atmosphere. Daytime temperatures are high throughout the year. Weather conditions are almost the same every day. The sun rises at about 6 a.m., and the early morning mists soon disperse. As the sun gets higher in the sky and temperatures rise, a lot of water evaporates from the rivers and forests. As the air rises it cools, and the water vapour condenses to produce great masses of cloud. By afternoon there is usually a torrential downpour, often with thunder and lightning. The clouds then break up, and at about 6 p.m. the sun sets and night begins. The regular evaporation and rainfall means that the weather is always humid and sticky as well as hot, and there are no summer and winter seasons as we know them.

Any climate with average monthly temperatures over 25°C and a rainfall of over 2 000 millimetres spread throughout the year will result in vegetation similar to that of Amazonia. The tallest trees rise

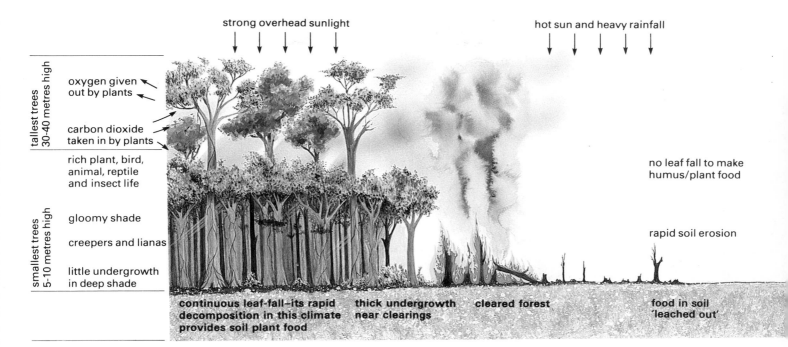

Natural rainforest and cleared rainforest

like giant pillars to over forty metres high, and are close enough together for their crowns to overlap. From above the trees seem to form a continuous canopy. Beneath this layer are other layers made by smaller trees, while there are masses of creepers and rope-like lianas looping and hanging from the trunks and branches. In places in the greeny gloom of the forest floor there are open spaces, but often there is a cover of bushes and undergrowth which is difficult to move through. Thousands of different sorts of birds, reptiles, insects and animals inhabit the jungle environment.

With the little change in climate there is no particular season for the trees to flower and fruit. Individual trees loose their leaves at different times, so the forest as a whole remains evergreen. The floor is covered by a thick, spongy layer of decaying leaves, and it is humus from these that provides the vegetation with food. Such rich plant growth suggests the soil is rich, but this is far from true. When the tree cover is removed, the heavy rainfall and heat soon wash plant food out of the thin reddish soils, and there is every danger of it being lost completely by erosion. There are other disastrous outcomes that we shall see in the next section.

1. Make a sketch from the photo of the rainforest on page 14. Label to show the daily movement of water and water vapour.
2. Plants grow and vegetation decays quickly in hot, humid climates. What happens to things left in hot, damp rooms? Can you think of any other everyday examples of chemical changes being speeded up by higher temperatures – in the kitchen or garden or rubbish bin, for example?
3. Why were forest areas *not* ruined by the 'slash and burn' methods and shifting cultivation of Indians in the past?
4. If soils are rapidly ruined or lost by forest clearing, why do you think rainforests are cut down or burned on a large scale?

The alligator is one of the less pleasant inhabitants of the tropical rainforest

2 Rainforest

Development or destruction?

Two of the men and their heavy machinery responsible for clearing the Amazon rainforest

'From the air, the Amazon forest is an endless dark green carpet with great brown rivers snaking their way across it. The signs of human activity – small clearings, curls of smoke, isolated townships and the thin bright red line of the Transamazon road ruled across it seem puny by comparison. Brazil feels it must conquer this immense wilderness for its resources and the living space it seems to offer.

Along the Belem to Brazilia road, towns are mushrooming where a dozen years ago there was only jungle. On either side of the road, broad swathes have been slashed and burned out of the forest and grassed over to pasture herds of white and brown zebu cattle. The road itself hums with timber lorries, cattle trucks, petrol tankers and giant earth moving vehicles . . . .

Maraba is one of these frontier townships accessible only by a gruelling sixteen-hour bus ride from Belem. Men have been coming here since the turn of the century, attracted by easy pickings first of natural rubber, then of diamonds. Then they came to collect brazil nuts from the majestic thirty-metre high "castanheiros" that tower over the forest's lower layers. And now they are driven by land hunger or greed for a share of the untapped mineral wealth that everyone assumes must lie here.'

The extract gives a slightly misleading impression of the nature of the changes taking place in the Amazon rainforest. In the past decade over 10 000 kilometres of road have been driven through the forest to

The forest burns

A new road cuts a broad scar through the jungle

enable it to be settled and developed. Large areas have been burned and over 300 ranches with more than six million head of cattle created in their place – with a great deal of financial help from the government. Another example is an enormous project at Jari near Manaus. Vast patches of rainforest are being replaced by faster-growing foreign tree species to feed a giant pulp mill. Over 35 000 people are employed on this huge enterprise, which is owned by an American multi-millionaire, Daniel K. Ludwig. Apart from these big ventures, the Brazilian government has resettled over a million colonists from poorer parts of the country. Many are peasant farmers struggling to make ends meet, but others now have quite prosperous farms growing crops of coffee, cocoa, palm oil and soya, especially along the more fertile river banks.

But there is a price to pay. It has been calculated that about one fifth of the rainforest has already been destroyed, and at the present rate of clearing there will be little left in twenty-five to thirty years time! This would not only be a waste of very valuable timber, but such a massive loss of forest could affect the oxygen and carbon dioxide balances in the atmosphere – with unknown but possibly disastrous results. Most of the soils are not rich, and without forest cover would soon deteriorate or be lost. The few remaining Indians will either be exterminated, find themselves on reserves or drift into urban slums.

Cattle graze in the new ranchlands of the Amazon

1. What does the extract tell us about *why* people from outside the area visited the forest **a)** in the past and **b)** in recent years?
2. Which words in the extract could be used as titles for photographs on these and the past few pages? Give four examples.
3. Look back to question 2 on page 17. How many trips between London and Edinburgh would be represented by the length of roads built through the rainforest in the past decade?
4. Describe two of the big changes taking place in Amazonia. For each of these say **a)** what is good about it **b)** who benefits from it **c)** what is bad about it (if anything!) and **d)** who suffers from it (if anyone!).

Most South American Indians have now been sucked into twentieth century society. Where can this man get the money to buy the goods on display in the shop?

2 Rainforest   World patterns

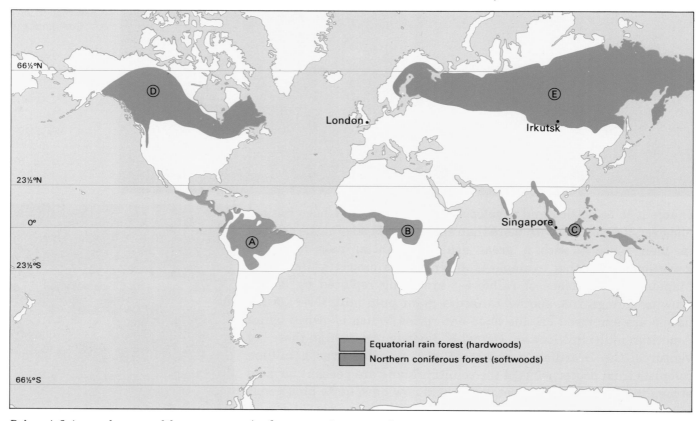

The world patterns of two different sorts of forest. These patterns are very generalised – within the shaded areas some of the forest may have been cleared

■ Equatorial rain forest (hardwoods)
■ Northern coniferous forest (softwoods)

Below: A flying snake – one of the many strange forms of wildlife in the tropical rainforest

A glance at the map shows us that there are other parts of the world where there are large areas with rainforests like that of Amazonia. This is not surprising when we remember that the forests are a result of a particular sort of climate – high temperatures and high rainfall in every month of the year – that is found throughout the world within a few degrees of latitude of the equator. There are two simple reasons why there is not an even band of rainforest circling the earth on either side of the equator (apart from the destruction by man). You can probably work these out if you look at a globe or atlas.

A few of the different varieties of equatorial rain forest are shown on these pages. It is nearly 40 000 kilometres around the equator, so it is hardly surprising to find a great deal of variety in the forests, depending on local circumstances. The photographs – and others in this book – also remind us that very valuable timbers such as mahogany and rosewood come from these forests. They also show that clearing in some places has been less ruthless than in Amazonia, and many crops are successfully grown on farms and plantations in these climates.

Even so, there is growing concern about all the worlds rainforests. It is very difficult to get accurate figures, but it is estimated that these forests have already been reduced to about sixty per cent of their original size. The Food and Agricultural Organisation has put the present annual rate of loss to be more than the area of England and Wales! The main threats are from logging, clearing for farming and settlement, and for mining. Clearly such massive change cannot go on for very much longer. Not only will the millions of people who live in and from the forests have to change their way of life. Just as important in the long run is the possible effect on soil, landscape and climate. There is ample evidence to suggest loss of trees on the present scale will lead to soil erosion, a wasteland of useless vegetation, widespread flooding, changes in rainfall pattern and perhaps even a change in the composition of the atmosphere. The forests and minerals in equatorial climates are valuable resources, but if they are wastefully or carelessly used they will be lost for ever.

1. From a globe or atlas find out the names of the five lines of latitude shown on the map. (Later on we will see what is special about them).
2. What are the names of the area or countries at A, B and C? How are climates at D and E likely to differ from those producing the rainforests? Which of the three graphs is for Singapore and which is for Irkutsk? Give reasons for your choice.
3. Write a few sentences describing the mangrove forests. What sort of feelings do you think you would have in this sort of forest just as the sun was setting?
4. Many areas of rainforest have been cleared and the land used to grow crops on a large scale (page 21). What are these large commercial farms called?
5. Why is it difficult to know how much forest is being cleared each year? Who decides what will happen to the forests? Who should decide what happens to the forests? Is it any of our business – or our responsibility?

Top: A mangrove swamp in Nigeria

Above: The chain saw has helped to speed up the clearance of the world's rainforests. Bendel State, Nigeria

Three climates: London, Singapore, Irkutsk. The temperature and rainfall graphs shown are averages. This means, for example, that the temperatures for any single month or for any single day, may be much higher or lower than the average

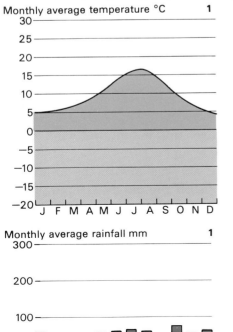
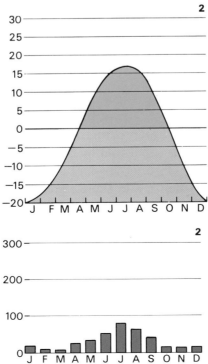
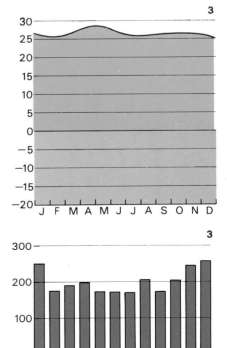

3 Desert   The Empty Quarter

A camel train

The camel remains the most important possession of the Bedouin

'The twin-engine Otter sputtered into life at Dhahran's busy airport. We taxied past three big Boeing 707's: Saudi Arabian Airlines, Middle East Airlines and Pakistan International Airlines. We were heading down to Saudi Arabia's Empty Quarter, the Rub' al-Khali, or "the Sands" as Bedouin call it. It is the largest uninterrupted mass of sand in the world. We headed south along the Arabian Gulf shore. The vivid coastal colours were faded by the dust storm we were going through. An hour later the air cleared as we headed inland from the Gulf. The sands near the coast were whitish in the glare of the sun, but soon turned pink in tone. We droned on. The waves became larger and larger, changing subtly from waves into mountains, sand mountains separated by broad white flats . . . .

On another occasion I had spent six weeks with the al-Murrah Bedouin and their camels on the edges of the Sands, sleeping in the desert, existing on camel's milk, dates, rice and – occasionally – camel meat. And, of course, the traditional strong Arab coffee and sweet tea.

Now, cruising through the grayish-blue sky, half asleep, my eyes fixed on something that looked at first like a cluster of bushes. We had spotted a well where hundreds of camels were being watered. From the south, long lines of camels glided toward the well, other long lines moved away to the north. As we flew closer we could see Bedouin hauling up water and others walking on with the herds. Two pickup trucks were parked near the well – a sign of the new prosperity of the country. Bedouin use the trucks nowadays to carry heavy gear such as tents, blankets and cooking utensils and perhaps very old or very young family members as well. In all the time I had spent living and travelling with Bedouin I had never seen so many camels on the move.'

Originally the desert proved such a harsh environment that nobody could live in it. Travelling through the desert and living in its interior only became possible when the wild camel had been caught and tamed. Together with his camels the Bedouin learned to survive in the desert lands. They became nomads, following the rains and the vegetation caused by them. It proved a struggle to survive, and produced groups or tribes of proud, tough people frequently fighting others to get use of the sparse vegetation and few wells in the desert – but traditionally very hospitable to strangers. They also traded with the people living in the villages of the desert border or in the oasis and wadis – river courses – in the desert. The Bedouin protected their cultivated lands in exchange for dates, rice and other crops from the fields. But as we have already noticed, changes have taken place in this very old way of life.

1. What clues do you get from the extract and the illustrations about a) the weather and b) the vegetation of the desert known as The Empty Quarter? What special line of latitude passes through the Arabian desert?
2. Using the information given, write a few sentences on 'the life of the Bedouin'. What would you like or dislike about it?
3. What signs are there in the extract and photograph that the old way of life is beginning to change?

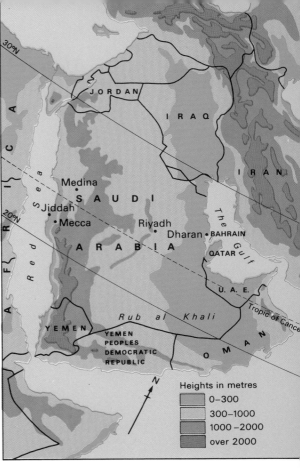

The Arabian Peninsula

A Bedouin settlement

The desert environment

Crescent-shaped sand dunes

Deserts are the driest places on earth. In deserts less than 25 centimetres of rain fall each year and it is rare to find surface water. In desert areas, though, such average figures mean little since rainfall varies so much. In some years there may be virtually none and in the next year ten times the average. It is the low totals, uncertainty of timing and varied amounts of rainfall that make desert environments so severe for plants and animals. There is so little rain because there is hardly any water vapour or clouds in the sky.

In most parts of the world clouds cut out and reflect a great deal of the sun's radiation, but in hot desert areas only about ten per cent is cut out. So the sun beats down out of a cloudless sky producing some of the highest temperatures in the world. On the other hand the cloudless sky means that heat is rapidly lost into space at night and it can become very cold.

This regular change from very hot days to cold nights causes any exposed rocks to expand and contract. The rock surface splinters or crumbles into stones and gravel and finally grains of sand. These materials are blown by the strong winds to wear away at other exposed rocks. This wind carving produces spectacular scenery in some places while elsewhere the sand forms great dunes, continually changing shape under the influence of the winds.

Plants need sunlight and food from the soil to grow. Lack of sunlight is no problem in the desert! But desert soils, unlike those in our climate, have little humus in them. This is because there is little plant or animal life to decay into the mostly sand and gravel soils. When it does rain the soils are quite fertile, however, because plant

An oasis in the desert. El-Oued in the Algerian Sahara

Tuareg women moving camp. Like the Bedouin the Tuareg are desert nomads. They are found mostly in the western Sahara

foods that are in the soil have not been washed out. Some desert plants have adapted themselves to take advantage of the infrequent rain very quickly before it evaporates by such means as developing long or horizontally spreading roots. Succulents and cacti have also adapted themselves to very dry conditions and to extremes of heat and cold. A very different sort of vegetation such as date palms is found in the occasional oasis or water hole where surface water can be reached by plant roots throughout the year. The water is carried by underground rocks from a long distance away where it rains more often.

The same environmental problems face desert animals. The camel is a good example of how they have adapted to the conditions. Long eyelashes, ear-flaps and muscular nostrils keep out the sand. The two-toed and webbed feet are ideal for walking in sand. Food is stored in the fat of the camel's hump, while its body heats and cools in such a way as to avoid too much loss of water by sweating. Camels can go very long distances without drinking water, and then recover quickly with a massive drinking bout. Camel fur traps air between the skin and air surface, so keeping the body cool. The clothing worn by the Bedouin provides similar protection. In the desert as in the rainforest, plants, animals and men have adapted to the environment.

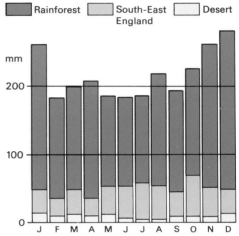

Desert rainfall compared with that of a rainforest area and South-East England

Below: How a 'barchan' sand dune moves downwind

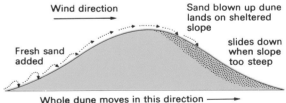

(a) View from the side

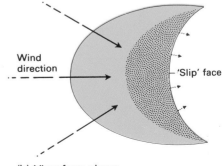

(b) View from above

1. Make a sketch of one of the camels and rider. Label to show as many features of the environment as possible, and the ways in which the camel and the rider have tried to adapt to it.
2. Look at the rainfall graphs. What are the similarities and differences between a) the amount of rainfall and b) the seasonal pattern of rainfall in the three places.
3. Draw a vertical line and divide it into thirteen equal parts. These represent temperature differences of 10°C. Start at the bottom of the line, and label −20°C, then −10°C and so on. You should end with the temperature of 100°C.
Now mark the following temperatures on the line:
a) Mid-day temperature in Libyan Sahara: 58°C. (the highest recorded air temperature). b) Ground temperatures under blazing desert sun: 93°C. c) Boiling point of water: 100°C. d) Night-time desert temperature of air: −10°C. e) The hottest temperature you have experienced. f) The coldest temperature you have experienced. g) The air temperature when you are doing this exercise.

3   Desert   Change in Saudi Arabia

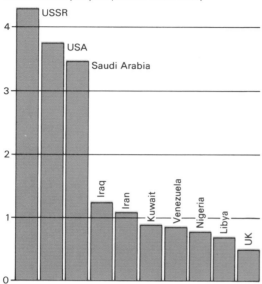

The world's main oil producers

Below right: Jiddah. A modern office block towers over the minaret of an ancient mosque

Below: A street scene in Riyadh

The largest country in the Arabian peninsula is the Kingdom of Saudi Arabia. This kingdom, ruled by the House of Sa'ud and its many princes, is nearly one third the size of the U.S.A., so it is not surprising that it contains some variety – rugged mountains and humid coastline as well as the great hot, dry desert.

The minaret of the mosque at Jiddah reminds us that this is not only an Arab, but also a Moslem country. Two of its cities – Mecca and Medina – are religious centres. Mohammed, the prophet who was responsible for the birth of the religion of Islam, was born in Mecca in AD 570. It is now a great centre of pilgrimage. The Arabs took the Islamic religion into many other countries of Africa, Europe and Asia, but eventually their empire collapsed and for a long time the people of Arabia were ruled by others. The Kingdom of Saudi Arabia was finally created under that name in 1932.

For a long time Saudi Arabia was a poor country because of its harsh environment, but many things have happened to this country recently. A great deal of this has been possible because of the discovery and development of oil. Saudi Arabia has become one of the world's leading producers of oil and exports it to many countries. The vast wealth this has earned has been used in a variety of ways, a few of which are shown here.

Until recently Haradh was little more than a water hole and a few general stores in the middle of an empty plain. Now it is the site of a new experimental farm run by the Ministry of Agriculture and Water to raise sheep and grow the grain to feed them. The aim is to raise some 150 000 sheep for their meat, hides and wool. As a by-product it also provides jobs for many Bedouin who want to change from their nomadic life of wandering the desert.

Newly irrigated fields at the Haradh experimental farm

Left: Oil tankers queue up to be loaded at Ras Tanura terminal

Symbol of Arab prosperity: burning off excess oil and gas

A great deal of the wealth has been used to build roads and to set up refineries, factories, plants to turn sea water into fresh water and to modernise and develop the urban centres. Many Saudi Arabians now live in blocks of flats such as the one shown here, and work in offices such as those in Riyadh. One of the most modern hospitals and clinical research centres in the world may be found in Riyadh. Many schools and colleges have also been built to educate and train young Saudis to gain new knowledge and skills. Saudi Arabia is a good example of a once poor country using new-found wealth to modernise itself while not sacrificing its past traditions and Islamic way of life for a 'westernised' one.

1. Copy the diagram of oil producers. Colour in the columns representing oil production from Arabia and the other neighbouring 'Middle East' countries in a colour different from the rest. Why are these countries so important in modern world affairs?
2. Look at the map on page 25. Measure the distance from Riyadh to Jiddah. (Compare with the approximate 500 km between London and Edinburgh)
3. How has the new wealth allowed the desert climate to be used at Haradh? What part of the climate has been 'provided' by the Saudi Arabians themselves? Near the coasts there are a number of desalination works. What are they for?
4. What news have you heard or pictures have you seen about Saudi Arabians that suggest that both individuals and the country as a whole are nowadays very wealthy?

3 Desert — World patterns

Monument Valley, Utah

A Kirghiz woman cooking. The Kirghiz live in a 'cold' desert on the Russian steppes, near the Afghan and Chinese borders

A general picture of the planetary wind system between the tropics. (The winds shown indicate the most frequent direction of movement)

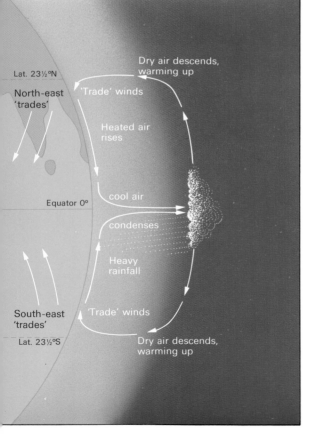

An area is usually called a desert if it receives less than about 25 centimetres of rainfall a year, although the hot sand deserts get far less than this. In comparison, south-east England receives about 60 centimetres. There is no sharp boundary where the annual rainfall rises above this figure, and true deserts merge into semi-deserts where there is a fair amount of scattered shrub vegetation.

Although the boundaries are hard to find, there is no doubt about where the hearts of the true deserts are. If we look at a map of the world, or better still, a globe, we notice something about their location – they are all in about the same latitude north and south of the equator. This gives us a clue about the reasons for this type of climate and environment.

Over the equator the air masses are heated and move upwards through the atmosphere, cooling and condensing to give the heavy daily rainfall already described in the section on rainforests. After reaching a certain height the air masses flatten out and begin to move towards the north and south. At about the latitude of the Tropic of Cancer and Tropic of Capricorn the air masses descend towards the earth's surface again. But now they are dry and as they get warmer they are able to hold and absorb more water vapour. As a result the air in these latitudes is dry, the sky is cloudless and the sun very hot. The air masses now move back towards the equator picking up moisture from seas, rivers, lakes and so on as it does so. (In fact, because the earth is spinning and producing a friction-like effect, the surface winds blow in a south-easterly and north-easterly direction, as shown in the diagram). Desert climates, then, are due to the great repeating circulations of air masses in the lower atmosphere.

There are many different types of desert and semi-desert landscapes, apart from that of the sand dunes. A few are shown here. In Monument valley, Utah, we can see all that is left from great layers of rock that once covered the whole area. The tall solid rock structures, known as mesas, buttes and pinnacles have been partially protected by caps of harder rock. But one day even these will have eroded away by temperature change, wind action, and the occasional storm. These desert landscapes are quite unlike the hills, valleys and plains found in rainy areas.

Altogether the deserts cover about fifteen per cent of the world's land surface. Until recently these harsh barren lands were almost useless to man. Recently, though, many areas have been made very fertile by using irrigation, while modern methods of exploration and mining have led to the discovery and production of great mineral wealth from the deserts.

1 Copy the diagram of air mass movement, adding the names 'Tropic of Cancer' and 'Tropic of Capricorn' in the correct place. Shade in the areas between these two lines of latitude and label as 'The Tropics.'
2 Copy the upper diagram of the earth from the illustration. Add a layer of shading around the earth to represent the atmosphere – not too thick! Now give *two* reasons why the mid-day sun is so hot at places on or near the Tropic of Cancer on 21 June. Why will places on or near the Tropic of Capricorn experience less heat from the mid-day sun on that day?
3 Look at the photograph of the aboriginal man. What suggests that this is semi-arid and not absolute desert? What are the signs that this aborigine is not completely dependent on the environment, but has contact with other people?
4 Write out the letters a–g. Use an atlas to find out the names of these hot desert areas and write them next to the correct letter.

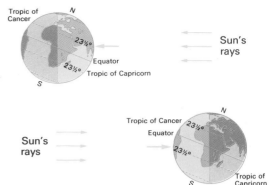

a) Midday sun overhead at the Tropic of Cancer on 21 June. Northern Hemisphere 'summer'. **b)** Midday sun overhead at Tropic of Capricorn on 21 December. Southern Hemisphere 'summer'

Some of the world's deserts contain great mineral wealth apart from oil. Blasting for diamonds in the Namib Desert, Southern Africa

An Australian aborigine. Most aborigines nowadays no longer hunt, but work on farms or in cities

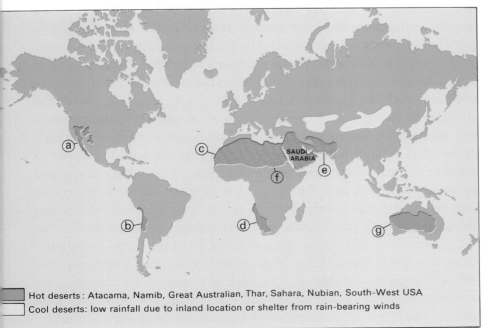

Hot deserts: Atacama, Namib, Great Australian, Thar, Sahara, Nubian, South-West USA
Cool deserts: low rainfall due to inland location or shelter from rain-bearing winds

Left: The world's deserts: areas of low rainfall

4 Monsoon lands — North Indian village

Tending the Hindu village shrine

The village shop. The goods may be modern, but the inside has changed little in the past hundred years

'Parsoiya is a collection of mud walls and thatch amid fields of wheat, maize, millet, sugar cane, potatoes, oil seeds and barley on the great North Indian plain. It is one of many thousands like it dotted across India's most populous state of Uttar Pradesh. It is just over fifty kilometres north of Lucknow, down a cart track off the main road. It has been there for at least 1 300 years, as can be seen from the stone images in the crumbling low-beamed temple named after the Hindu god Thakurdwara, which stands on a hillock overlooking the village.'

The population of the village reached 710 in 1977, but a year later it had dropped to 542, largely due to landless families leaving to try and find work in Lucknow, Delhi and elsewhere. The life of the villagers is dominated by the need to produce enough food from crops and animals to survive. Farming is still a matter of hand and bullock power. The village has twelve wells from which drinking water is drawn and one overworked irrigation canal – electricity and piped water are both a dream for the future. Because of this the farming year is still largely determined by the seasons, especially the changing rainfall. Land has to be ploughed and prepared during the hot dry months before June, a time of cloudless skies and great heat. Everyone waits impatiently and anxiously for the summer rains. When they come – if they do – rainfall is torrential for a few weeks, providing essential water for the soil and plants. The rains then ease off, and the temperatures fall until the annual cycle begins again.

'Life in Parsoiya has much to do with the complications of caste, the eking out of a subsistence from a hectare or two of land, or a daily income of a few rupees for hours of labour in the hot sun. Like other local leaders and teachers, Bhagwan-Deen Gupta, deputy headman of the village is a member of the upper castes who also hold by far the biggest share of the 240 hectares of village land. A number of the Harijan families – Harijans are the scheduled, outcast or so-called untouchable caste – appear to hold several hectares of land, following recent efforts at land reform, but some have none and are reliant on keeping a few cows or pigs, or on labouring work. Some new high yielding strains of wheat and rice are being grown, but only by a few of the better-off farmers. The rest find it difficult to raise loans to buy fertilizers and seed. Even irrigation cannot be used because of lack of money for pumps. The result is that the village has not moved far from subsistence farming relying on milk from cows, buffaloes and goats and home grown vegetables to eke out the diet. According to the deputy headman, for nine months of the year many families manage on one meal a day, frequently making do with a wheat "roti" flavoured with chillies and salt.'

1. Look at the map on page 34. What is the name of the great river that drains the huge plain in North India? What is the name of the other long river, rising in China, that joins it at its delta? What is the name of the sea into which both rivers flow?
2. Copy the rainfall graph on page 35 and label to show these three seasons: hot and dry; hot and wet; cool dry. Show when you think ploughing the soil, planting seeds and harvesting the summer crops takes place.
3. Choose one of the scenes in this section, sketch it and then label to show signs that the farmers or villagers are fairly poor. Suggest two or three improvements they might make if they had the money to do so.
4. The caste system by which everyone is born to do certain jobs and live in certain ways, without ever changing, is part of Hindu belief. Strict caste behaviour has broken down in some towns and cities, but it is still often strong in many villages. What are the signs of caste in Parsoiya? What do you think might be good and what bad about the caste system?

The bullock cart is still the main means of transport in a village like Parsoiya

Winnowing – separating the wheat from the chaff. In the background are the typical mud buildings of a North Indian village

4 Monsoon lands

The burst of the monsoon

A monsoon rain cloud hovers over a West Bengal rice field

The arrival of the summer rains, known as the monsoon, is an event that has happened since the beginning of recorded time – apart from disastrous years when for some reason they fail to do so. The monsoon 'bursts' at different times throughout India, as the map shows. This happens first in the south, then the centre and north-east and finally the north-west. The dates are the usual ones – but the rains may be early, late or sometimes never occur at all.

Because of this uncertainty, and because the failure of the monsoon can mean starvation or death, the period leading up to the burst is a time of great anxiety. These feelings are well described in this extract:

'With the passing of each week the tension grew. It was not only the awful heat that wore nerves finer and finer, but terror as well, the terror of famine and disease and the horror of that burning sun which nerves could bear no longer. The storm, accompanied by a sudden fierce wind, came up quickly, covering all the stars that were like the diamonds of the Maharani as if a thick curtain had been drawn across

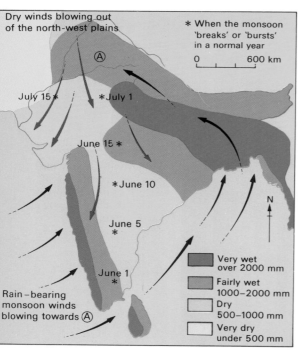

The usual dates of the arrival of the monsoon at different parts of India

Right: India

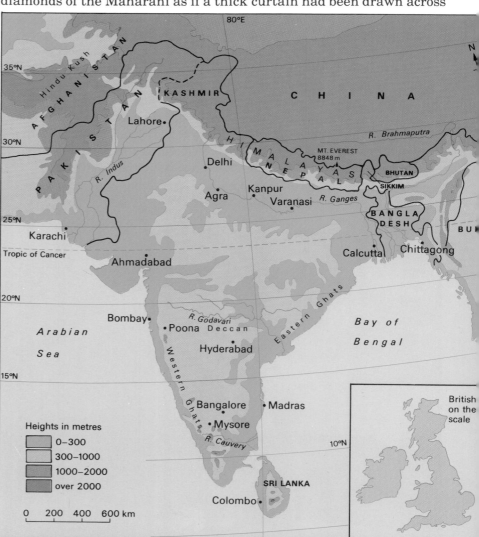

Transplanting rice seedlings into a paddy field. Farmers in India depend on the monsoon to water and irrigate their crops

them, and the thunder and wild flashes of lightning drove the gigantic bats into new and wild flutterings above the tank or pond ... great drops began to splatter in the thick dust. The branches of the mango trees whipped black against the wild glare of lightning, and the water fell in torrents on the parched thirsty earth. Tomorrow it would be green again, miraculously green with the miracle of the monsoon.'

As the maps show, the warm rain-bearing winds come from the Indian ocean to the south. Some move over the Arabian Sea before crossing the coast and rising over the Western Ghats. Others swing across the Bay of Bengal, meet the mass of the Himalayan mountains and then swing again up the Ganges valley and over the great plain of north India. The north-west of India has become very hot during May and June, and low pressure has encouraged this particular pattern of winds. During the cooler part of the year the wind pattern is reversed and India is crossed by cool, dry winds that bring little rain.

The amount of rain that falls varies from place to place. The rainier areas were once covered by open, deciduous woodland, with trees shedding their leaves during the hot dry season (not during the cold winter, as in Britain). Most have now been cleared for farming, apart from the steep mountains. The drier parts are covered with thorny bushes, scrub and grass that shrivel up in the hot, dry season.

1. Why does the south of India get the monsoon rains before the north? Why is rainfall along the west coast so much more than in the interior? Why does the monsoon rain fall such a long way inland from the mouth of the Ganges and the Bay of Bengal?
2. Give two differences between both the rainfall and the temperatures of Calcutta and London.
3. From the map of the monsoon draw two outlines of India the same size. On one show the broad pattern of air movements in June and in the second the air movements in December. Carefully title each map.
4. Describe the scene of people in the fields. What do you think you would enjoy or dislike about living and working in this place?

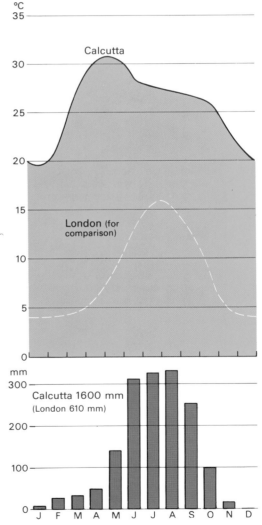

The temperature and rainfall graphs of Calcutta. What happens to the temperature when the monsoon rains 'burst'?

4 Monsoon lands

Flood, cyclone and drought: a year in India

A press photograph, sent by 'laserphoto', of a flood in eastern India with the message that came with it

Above right: A cyclone in full blast

This ship has been carried on to dry land by the force of a cyclone

The last few pages have shown what the monsoon climate is like under average or normal conditions, but so often it is one extreme or another that affects the land and the lives of millions of people. The following reports are of three extremes of weather in India that occurred in just over one year.

September 1978

'One in every twenty among India's population of six hundred million is now a flood victim. Since the onset of the annual monsoon scourge over northern India in late June there have been 898 known deaths, more than 46 000 villages have been inundated, 600 000 homes have been swept away or badly damaged and 4 000 head of cattle lost. From Old Delhi's Red Fort the flooded Yamuna river merges indistinguishably into a sheet of water stretching without interruption for about eight kilometres to the east and sixteen to the north. This is the extent of this years' monsoon flooding around the Indian capital.'

May 1979

'A cyclone in India's southern Andhra Pradesh state last week killed nearly 600 people and made more than 1.2 million homeless. India's East Coast suffered less than feared. Coastal villagers in Andhra Pradesh were evacuated thanks to several days' warning – the cyclone veered around in the Bay of Bengal a week before striking 150 kilometres of coast at 4 a.m. on Saturday May 12th. The cyclone of November 1977 killed more than 10 000 people. The exact total has never been established as many were migrant farm workers, brought in at the start of the monsoon.'

November 1979
'A journey through several of Uttar Pradesh's worst affected districts from Varanasi to Kanpur has given me the general impression that India's general election is unwanted by millions of rural people enduring the most devastating drought since Independence in 1947. Over northern India as a whole there now exists a hungry rural mass of twenty million people, half of them children. The harsh realities of rural India are highlighted by the drought. You travel for miles by road seeing the encrusted barren brown earth and then the rare patches of bright green paddy, patches belonging to farmers able to afford their own tubewells. With an officially estimated loss of foodgrains from the monsoon crop of eleven million tonnes – compared with a total of seventy-eight million tonnes in 1978 – India's Agricultural Secretary is hoping to recoup on the winter "rabi" crop. India's winter crop, though, is overwhelmingly produced under irrigation; but less than thirty per cent of the country's agricultural land is irrigated.'

A victim of drought – Bihar, India. Note the parched earth

1. What was the type of damage or disaster caused by these three events? Can you think of similar extremes of climate leading to damage in this country in recent years? If so, how many people were involved, or killed and how much damage was done?
2. Imagine you are a newspaper reporter sent to India to write about the monsoon and its effects. Choose one of the photographs and write a short newspaper report about it.
3. Why is India's 'winter' crop so dependent on irrigation? If irrigation is so important why is only 30 per cent of the land irrigated?
4. Why do people go on living in places likely to suffer from flood, cyclones or drought?
5. What do you notice about the latitudes where most of these strong winds occur?

Collecting water from a well, using a modern hand pump

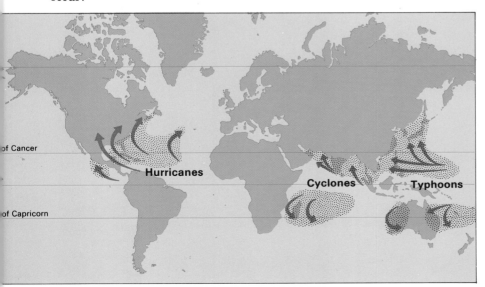

The world distribution of hurricanes, cyclones and typhoons

4 Monsoon lands

Cities of the northern plain

Delhi. This picture is of the 'old' town as opposed to New Delhi where India's Houses of Parliament may be found

In the far north of India is the great mountain chain of the Himalayas. In the centre is the lower, flatter but still quite high plateau known as the Deccan. Between the mountains and the plateau is the large plain drained by the Ganges and its tributaries and influenced by the monsoon climate. A large proportion of India's 600 million people live on the plain, mostly in the thousands of villages and small towns like Parsoiya. About eighty per cent of India's population live in rural areas, but because the total population is so large this means that the number of city dwellers is more than the total population of Britain! There are also more large cities than in Britain – and their size is rapidly increasing.

One of the biggest cities of the plain is Delhi. Nearby is the newer city of New Delhi, capital of the country. India is a poor country and there is no question that millions of people live in dreadful conditions. But it is important to realise that Indian cities also have modern buildings, banks, offices, shops, factories and transport systems. There are also many very wealthy people, and the contrast between the wealthy and poor and between the educated and uneducated is greater than in Britain. We may be shocked by the poverty, but can also admire the skill of many Indians and the colour and beauty of much of their art, crafts and buildings.

India is a country where Hinduism is the main religion. The beliefs and practices of Hindus are very different from Christianity, the main religion in Britain (There are many Christians in India and

The cities of the North Indian plain

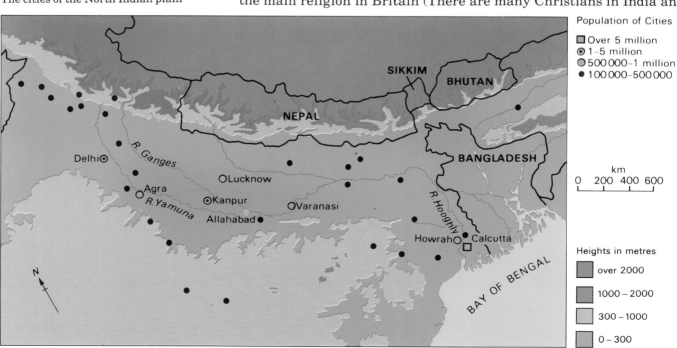

Hindus in Britain, of course). The importance of the caste system and of the village temple has already been mentioned, and on these pages we can see other signs of Hinduism. Cattle are revered in the Hindu religion and Hindus are forbidden to eat their flesh. This prevents the growth of beef farming and results in their being allowed to wander around urban areas. The Ganges is one of several sacred rivers and the steps leading down to the river from the large and ancient temples at Varanasi are usually thronged with worshippers bathing in its waters.

Even in the cities, water is a great problem. Calcutta is thought to have over six million inhabitants, many of whom have come to live there from other parts of India or Bangladesh. All the services such as housing, transport, health and education are under strain. In spite of its site on the banks of the Hooghly river, a branch of the Ganges in the delta area near its mouth, drainage and water supply has always been a problem. Water is in short supply, so every possible source is used. Filtered water is available from thousands of fountains throughout the city, but many people have to use untreated water from the river, or supplies that may be contaminated by faulty drainage. In cases like this there is always a danger of water-borne diseases such as cholera and typhoid.

On the edge of the sacred River Ganges, at Varanasi (or Benares), Hindus bathe themselves as an act of piety

Two faces of Calcutta: the new underground railway under construction and, in the background, the Victoria Memorial, built by the British

1. Which cities and rivers have been mentioned in the section on the monsoon lands (six cities, three rivers)?
2. Copy the key for the sizes of the cities of the northern plain. By the side of each size write the number of cities in that category or group. What is the difference in number and the average distance apart of cities over 1 million and less than 500 000?
3. Compare the locations and sites of Calcutta and Allahabad.
4. Choose one of the urban scenes and describe the differences and similarities between it and a British town or city scene.

Cows and overloaded trams are both a common sight in the streets of Calcutta. The Hooghly bridge can be seen in the background

5 Polar lands

Eskimo whale hunt

As soon as a whale is spotted, the Eskimo crew (or Inuit, as they prefer to be known) paddles out to meet it (*top*). Once caught, the whale is cut up and shared around (*right*). Eskimos also hunt seals from the shore with high velocity rifles (*above*)

Each spring the Eskimos of north-west Alaska hunt the big, slow-swimming Bowhead whale. They have done this for thousands of years. Throughout April they wait for the frozen sea to melt. Then the whales will begin to swim along the open channels to their summer feeding grounds in the Arctic Ocean.

Each crew of seven or eight Eskimos sets up a camp on the edge of the ice. Near the sealskin boat there are canvas wind-breaks. There is also a tent where they eat their hot meals of rice, duck, spaghetti and stew. They may have to wait for days before a whale is spotted. It can be bitterly cold, and sometimes the wind is so strong the crews are driven back to the shelter of their village.

When a whale is spotted the light sealskin boat is slid into the water and paddled quietly towards it. As the broad, black back rises out of the water explosive harpoons are fired or thrown into the whale. As it frantically plunges deep and surfaces to escape, more harpoons are fired into it until the whale is killed or dies from exhaustion. The dead whale is then hauled ashore and cut up to be shared out among the Eskimo community.

Some wounded whales dive under the ice and escape. Others are never spotted and make their way into the safe, deep waters of the Arctic for the summer. In autumn they begin the return journey southwards before the sea freezes over for the winter.

For hundreds of years whales have been ruthlessly hunted by whaling ships for their meat, oil and other products. As a result many species are in danger of complete extinction. Whales are of two distinct sorts. Those of the baleen group have bony plates fixed to their jaws to filter plankton and fish from seawater. The Bowheads are like this. The other sort are toothed, and these include dolphins, porpoises and killer whales. Whales are intelligent animals, well adapted to an ocean environment. They can dive to enormous depths without harm. They are able to migrate thousands of kilometres in search of food and can navigate by echo in total darkness. They can 'talk' to each other through a sort of song. It would be tragic and inexcusable if human beings exterminated these intelligent sea animals that have existed for some forty million years.

Many countries belong to the International Whaling Commission that sets a 'quota' or limit on the number of whales that can be caught each year. Only a small number of Bowheads can now be caught. A few years ago the Point Hope Eskimos caught about ten a year. Now they are allowed only two!

1 What are the signs from the illustrations that this was a fairly recent whale hunt?
2 Imagine you are in the sealskin boat, or cutting up the whale. Describe your feelings.
3 The Eskimos have hunted whales for centuries, but now they are being forbidden to do so. What do you think are the reasons for and against limiting their catch to a few each year?
4 Make a list of all the things you have read or seen in this section that tell us that it is very cold in these places near the Arctic Ocean.
5 The whales migrate south in July before the seas ice over. **a)** What does 'migrate' mean? **b)** Why does it get warm enough in the Arctic summer for some sea ice to melt? **c)** Why doesn't it get as hot as near the equator?

A Bowhead whale, compared in size with a man

Whale catch quotas

	1978/79	1979/80
Bowhead	18	18
Total	19 722	15 883

The migration route of the Bowhead whale

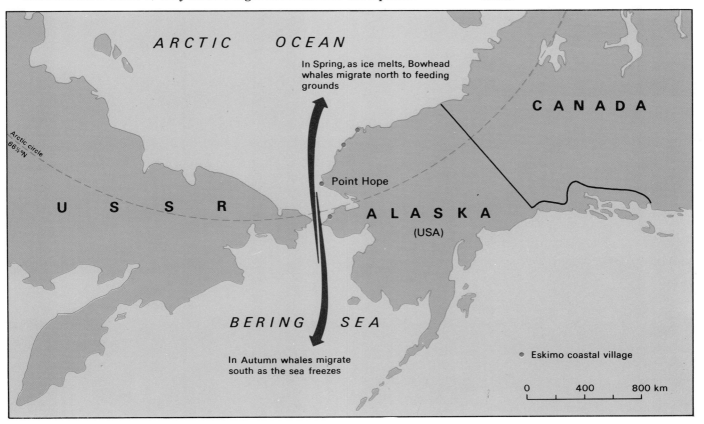

5 Polar lands

The Arctic

On the way to the North Pole: Wally Herbert, British explorer

The far from friendly adult polar bear

The northernmost part of our planet consists of a huge frozen ocean surrounded by the continental land masses of North America, Europe and Asia. The ice covering the ocean is known as pack ice, and in the photograph we can see a team of husky dogs hauling a sledge over the pack ice on a solo journey across these cold, barren wastelands. Greenland is the only land mass in these northerly or 'high' latitudes and it is covered by vast ice sheets.

In the northern summer, the edges of the frozen sea melt and turn into drift ice and the pack ice shrinks in size to expose some of the Arctic Ocean. The snow on the land melts and the top layer of soil may thaw out for a month or two. These lands are known as the tundra, and they merge southwards into the northern forests. The two photographs of Fort Franklin show the differences between the winter and summer seasons in these lands on the southern fringes of the Arctic.

A glance at the diagrams on these two pages explains why these high latitudes are so cold. North of the latitude known as the Arctic Circle, there are times in the winter when the sun never even rises above the horizon. At the North Pole itself, this lasts for about six

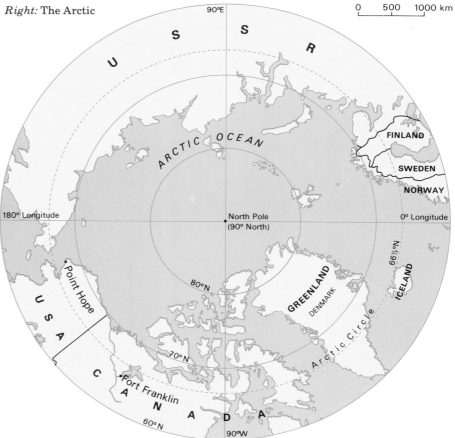

Right: The Arctic

42

months. In the summer, when the North Pole is tilted towards the sun there are periods when the sun does not set, and the Pole itself receives six months of daylight. It is easy to understand why it is bitterly cold without any daylight or sunshine, but harder to explain the low temperatures when the sun doesn't set for weeks on end. The reason is that the sun remains low in the sky even at midday, so it never has a strong heating effect.

The long, dark winters and cold weather make life very difficult, but plants, animals and man have adapted themselves to these harsh conditions. Two of the largest animals of the tundra are the wild caribou, or reindeer as the domesticated animal in Europe is called, and the musk-ox. The caribou are plant eaters, depending on a lichen that they find by scraping away the snow. The musk-ox has a tremendously thick coat protecting it from the savage winds. Another large creature is the polar bear. Its coat protects it from the cold while its hairy soles give it a good grip on the ice. The polar bear also eats plants, but feeds on seal, fish and birds as well. It is a remarkably good swimmer, and has been seen as far as 300 kilometres from land!

Apart from these there is a great range of smaller land animals, sea creatures such as whales, seals and walruses and birds ranging from the puffin to the fulmar. And in the short summer there are the flies and mosquitos that are a terrible nuisance to everybody. Yet they are part of the 'ecosystem' and 'web of life' of the place. The Eskimos and Indians too, have adapted to the environment and show remarkable skill and intelligence in using every possible resource.

1 Using your atlas give the approximate latitudes of **a**) Point Hope, **b**) Fort Franklin, **c**) the northernmost tip of mainland Europe, **d**) the northernmost tip of mainland Asia. For each of these four locations name the country they are in.
2 Copy the two diagrams of the Arctic landscape, but shade them or draw the position of the sun to illustrate 'midnight, December' and 'midday, June'.
3 What has drift ice to do with the sinking of the ocean liner 'Titanic'?
4 Write about the ways in which the Eskimos used the wildlife of the Arctic lands for their shelter, clothing, food and equipment. Why was there no danger of Eskimos exterminating wildlife in the past?

Fort Franklin in early summer (*top*) and in winter (*above*)

The North Pole at midnight in June and at midday in December

midnight – June

midday – December

5 Polar lands

Change in the Arctic

Right: An Eskimo (Inuit) family in a modern house

The church at Fort Franklin has been built in the shape of an Indian tipi

A geography lesson for the Grade 5 (12 year olds) at Fort Franklin school

Fort Franklin is a small Indian settlement on the shores of the huge Great Bear Lake in northern Canada. Nearby is the Bear River that drains the lake into the Mackenzie River which in turn flows into the Arctic Ocean. Until about twenty years ago most of the Indian families were nomadic, living in tents, hunting moose and caribou, fishing for trout and herring and trapping mink, marten and beaver for their skins. Nowadays, over 300 Indians together with about a dozen white Canadians and 400 dogs live in the fifty or so log cabins in the village.

The Indians still get most of their needs from the surrounding area. After all the nearest settlement is another village about 150 kilometres to the west and the nearest railway almost 1 000 kilometres away! Recently, with the help of the Government, they have built a church, village hall, nursing station and two stores. Money is earned by selling furs to the local Hudson's Bay store, handicrafts made from local materials and by acting as guides to the small number of summer visitors. There are about 6 000 Indians in northern Canada. Unlike the Eskimos they kept to the southern edge of the treeless barren lands of the Arctic. Nevertheless the settlements are still very isolated, and the Indians are only just beginning to be affected by the outside world.

We have already seen how the Eskimos of Point Hope have been affected by others. Not only have they obtained modern weapons and equipment for catching the Bowhead whales, but the number they are allowed to catch has been decided for them. But much more far-reaching changes are about to sweep the whole of north Alaska. Already oil is produced at Prudhoe Bay and piped across the mountains and valleys southwards to Valdez oil terminal. Exploration for oil and gas is taking place onshore and out in the open sea – when the ships can break the ice! The Eskimos fear the effects this might have on the feeding grounds of the whales – the risk of pollution in these icy wastes is high. But this is only a part of the story. It is almost certain that there are vast deposits of oil, gas, coal, and many valuable minerals in Alaska. But it is one of the few remaining wilderness areas in the world. So the Alaskans and Americans as a whole are struggling to decide how far they can exploit these vital resources without destroying or harming the wildlife and environment that has slowly evolved since the world began.

In the meantime the lives of the Point Hope Eskimos have greatly changed. They live in frame-built houses, not tents or snow igloos. Electricity is used, and some families own TV sets, hi-fi equipment and even a deep freezer – a lot of food is now imported! Eskimos can now earn a salary as workers for oil companies or for the government at military camps and as carpenters and electricians. Although the whale hunt goes on, life is certainly easier – and very different.

1 What are the approximate latitudes of the northern and southern ends of the Trans-Alaskan pipeline? Compare this distance with a similar one in Britain – if its northern end were in the Shetland Islands, where would its southern end be?
2 Look again at the map below. Apart from the cold, what other problems are there in piping oil across Alaska?
3 Choose either the Indian village at Fort Franklin or the Eskimo village at Point Hope. Imagine you lived there. Describe your feelings about the changes that have taken place.
4 Write down the arguments for and against developing the resources of the Alaskan wilderness. Would you be prepared to pay a little more for the wealth from the area if it meant protecting the environment and wildlife?

These men are sawing up a frozen caribou. The boy seems pleased with his cut!

Below: Alaskan oil has brought work to many Eskimos, helping them to acquire mechanical skills

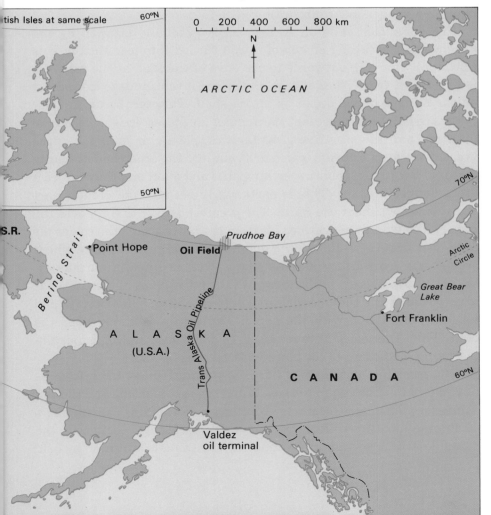

Left: Alaska

5 Polar lands

Antarctica

Right: An Antarctic research ship ploughing through the pack ice

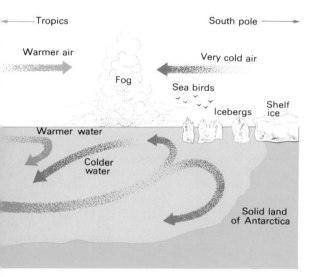

The edge of the ice-sheet, where warm and cold currents meet

The Antarctic ice shelf

Unlike the Arctic, which is frozen ocean surrounded by land, Antarctica is a continent surrounded by oceans. It is very large – far bigger than the United States of America. Nearly all this land mass, except the very tops of the highest mountain peaks, is covered with a thick layer of ice that averages over 1 800 metres. Ben Nevis, the highest peak in Britain is only 1 343 metres above sea level. This gives some idea of the massive thickness of ice and height of the mountain ranges in Antarctica.

The snow that falls on Antarctica is gradually crushed by the weight of snow falling on top of it and turned into ice. Most of it is continually moving outwards to the edge of the land at the rate of a few metres a year. When it reaches the sea it continues to extend out on the surface of the ocean as an ice shelf. This can be up to sixty metres in height, and present spectacular cliff faces. Ice shelves make up about half the coastline of Antarctica. Sooner or later the edges break off to form icebergs which drift out into the ocean before finally melting. Because of their uncertain paths and giant size they can be a great danger to ships. This is particularly because the greater part lies submerged under the water. The greatest danger to ships is in the zone of fogs where the warmer air from the north meets the colder air from over the Antarctic. As the diagram shows, this is a zone where ocean water from different areas mix, as well as different air masses.

There are several similarities with the polar lands of the Arctic. It is very cold, while there is the same pattern of long summer days and long winter nights, with the sun never setting or rising near the South Pole. The seasons are reversed, though. The Antarctic midsummer of midnight sun takes place in December.

The most common form of life is the minute plant life known as phytoplankton floating in the sea. When the ice breaks up and the sun shines for long hours this phytoplankton grows rapidly. Vast numbers of minute creatures, feed on the drifting plants, including the valuable krill, a shrimp-like animal. These animals are eaten by fish,

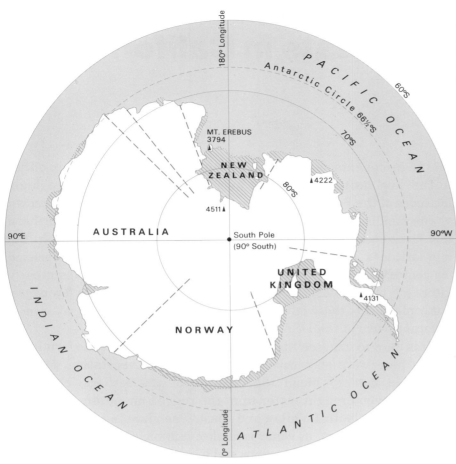

Antarctica. The major powers and countries with territorial claims in Antarctica (Argentina, Australia, Chile, France, New Zealand, Norway and the United Kingdom) agreed in 1959 not to press their claims during the 30 years up to 1989

whalebone whales, birds and seals. These larger animals, birds and fish are eaten in turn by leopard seals and the killer whales. The penguins that live in the Antarctic also feed on krill, fish and squid. They spend most of the year at sea, then come ashore to nest in great colonies near the shore – sometimes as many as a quarter of a million packed together.

Although Antarctica began to be explored over 200 years ago, it is only in this century that detailed exploration and scientific investigations began. Twelve nations now have more or less permanent research stations in the Antarctic, and they have signed treaties to share their results. Antarctica contains vast wealth and it will be important for nations to agree peacefully about its ownership.

1. You have noticed the lines of latitude separating the northern and southern parts of the British Isles – roughly about 10° of latitude apart. How many of these distances are there between the edge of the map above and the South Pole?
2. Draw a diagram to show the link between energy from the sun, the food chain in the Antarctic lands and man – remember that men hunt seals and whales. If so many creatures feed on others, why hasn't all life been exterminated before now?
3. The four largest areas of Antarctica shown on the map are claimed by Norway, Britain, Australia and New Zealand. Why do you think these countries have claimed the territories?
4. Why has Antarctica been one of the last places to be explored and investigated? Should the resources of Antarctica be shared between all nations? If not, why not?

When penguins lay their eggs they like to do it in company!

6 Mid-latitudes — Contrasts in California

The environments and ways of life considered so far have been either in the 'low' latitudes on either side of the equator, or in the 'high' latitudes of the Polar areas. Between these are two zones often known as the middle or 'mid-' latitudes. Britain is included in the northern mid-latitude zone. There are many different sorts of climate in the mid-latitudes. Apart from this, large parts of them have been settled for centuries, and the vegetation cleared and the landscape greatly altered. It is impossible to describe all these differences in a few pages, but some idea of these environments can be had by looking at several different parts of the USA.

It is hard to imagine that parts of California such as the farmlands shown here were once covered with dry scrub vegetation. Modern California is very much the creation of the settlers, especially engineers who brought water from the rugged mountains to the dry plains of the Central Valley and coast. The climate is excellent for crop-growing, as long as water is made available. Californian farmers get their income from many crops and livestock products, but the state is best known for its fruits and vegetables. These are not grown everywhere, as the map shows. Areas specialise, and crops are concentrated on the most suitable high-yielding land. But almost all depend on irrigation. As a result there is more irrigated farmland in California than in any other state.

It would be wrong to think of all California as a huge irrigated farm, though. It is much wetter in the north and in the various

An irrigated field in California. The farmer floods each channel running between the plants

Below right: Californian fruit crops

Below: California

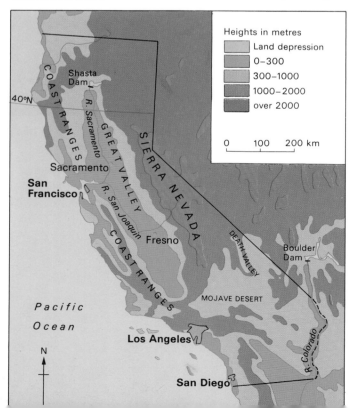

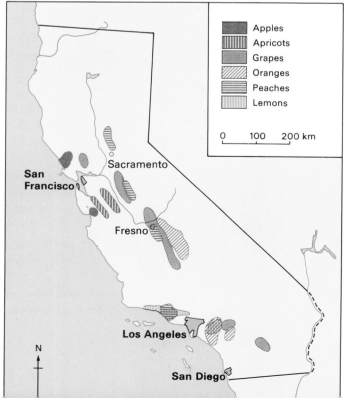

The giant Sequoia trees of the Sierra Nevada

Left: Irrigated farmland near Los Angeles

mountain ranges, and very large trees grow in the forests of the Coastal Ranges. The state has a very varied relief and covers a wide range of latitude – and both affect the climate.

Settlers have been attracted to the West Coast for many reasons other than farming. Gold was the big pull in the middle of the last century. The goldminers were known as 'Fortyniners'. When the railways reached the state in 1869 many millions of people migrated there to the cities and farms.

The early film industry was based there because of the climate and millions are still attracted there by the sunshine. Many things are made in the factories, ranging from ships and motor cars to aeroplanes, aerospace and microelectronics products. California, with its two big cities of Los Angeles and San Francisco, now has the largest population of any American state.

1. Look at a map of the World and say why it makes more sense to illustrate mid-latitude environments from the northern rather than from the southern hemisphere.
2. Look at the map of relief of land in California. Imagine you start at the coast 200 kilometres north of San Francisco and walk eastwards across the state. Draw a profile or cross section of your route, showing the way the land rises and falls. Label the names of the hills, valleys and rivers.
3. **a**) Why should the northern coast of California be a little cooler at all seasons than the southern coasts near San Diego? **b**) Why should the Coast Ranges be rainier than the Central Valley?
4. Make a sketch of the irrigated farmland to show how the land is used in different parts of the area.
5. What are the signs that Los Angeles is a rapidly growing, modern city in one of the most prosperous countries in the world?

Los Angeles. California contains some of the biggest cities in the United States. Los Angeles suggests what cities dominated by motor cars could be like

6 Mid-latitudes

Mountain, plateau and plain

Right: Sunset over the Grand Canyon

The Boulder Dam on the Colorado river at the Nevada–Arizona border

A cross-section from California to the Great Plains

It is still another 1 800 kilometres from the Sierra Nevada mountains and the state boundary of California to the centre of the USA. The first two-thirds of this – a distance more than that from London to Berlin – consists of a series of mountains and plateaux. Some idea of the changing relief of the area is given in the diagram, but it is difficult from this to get a real impression of the spectacular scenery or the variety of climate and weather.

The Colorado Plateau which covers much of northern Arizona and southern Utah is made of sedimentary rocks. They are almost as horizontal now as when they were formed, but they have been raised so high that the surface of the plateau is about 2 500 metres above sea level (almost twice as high as the top of Ben Nevis, the highest point in Britain!). A few rivers rise in the mountains to the east and flow across the plateau. The Colorado river is the biggest of these. It has cut a deep steep-sided gorge into the sedimentary rocks, known as the Grand Canyon. The Grand Canyon is 200 kilometres long and about 1.6 kilometres deep. There has been little erosion of the sides of the gorge because the area is very dry and little rain falls. The sides of the gorge form a series of steps due to the differences in hardness of the rocks. Further downstream the Colorado river has been dammed by the giant Boulder Dam, creating Lake Mead. The water is used for irrigation, hydro-electric power and domestic needs in the towns.

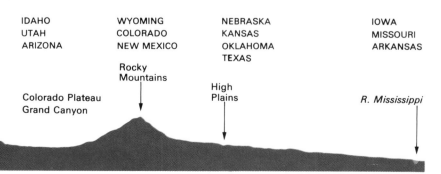

50

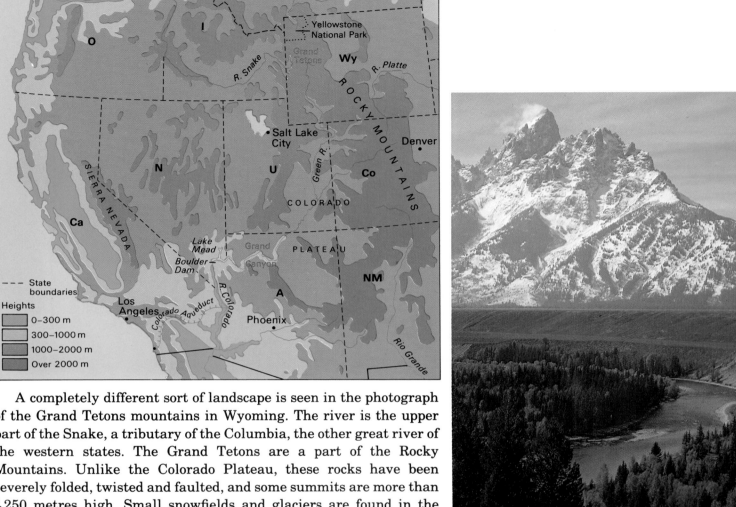

The mountain States of the USA

The Grand Tetons in Wyoming. In the foreground is the Snake River in its upper course near its source

A completely different sort of landscape is seen in the photograph of the Grand Tetons mountains in Wyoming. The river is the upper part of the Snake, a tributary of the Columbia, the other great river of the western states. The Grand Tetons are a part of the Rocky Mountains. Unlike the Colorado Plateau, these rocks have been severely folded, twisted and faulted, and some summits are more than 4 250 metres high. Small snowfields and glaciers are found in the highest parts, and the land has been shaped by ice sheets and glaciers in the past. As we can see, the lower slopes are covered by mountain pasture and below that a zone of coniferous forests. It is not surprising that many parts of the Rockies are important tourist areas. Near the Grand Tetons is the world-famous Yellowstone National Park.

From the map we can see that some very large rivers rise in the Rockies and flow eastwards. They cross first the high plains and then the low plains in middle America.

1 If you went in a straight line from the Sierra Nevada mountains to the city of Denver, through which four states would you have passed? What three rivers rising near the Grand Tetons flow in west, south and north-east directions?
2 Choose one of the two scenes (Grand Canyon or Grand Tetons) and make a labelled sketch to point out some of the main features that have been mentioned in the description or that you think worth emphasising.
3 Describe your feelings about the landscape shown in either of the two scenes.
4 What are the reasons for turning certain areas into National Parks? Are there any dangers in doing this?

6 Mid-latitudes The Great Plains

Above: An early photograph of a Plains Indian. The children are sitting on a 'travois' which, without knowledge of the wheel, was used as a means of transport

Right: An early European homesteader in Nebraska. With little wood available, turf was used as a building material

A storm in the 'dust-bowl' in the 1930s. Poor farming led to the destruction of the top soil and to dust-storms like this

Some of the rivers that rise in the Rockies such as the Platte flow eastwards across the Great Plains to join the Mississipi that drains the middle of America. The Plains near the Rockies are fairly level but at a height of about 1 500 metres above sea level – Denver, the main city, is sometimes called 'the mile-high city'! From here the land descends for hundreds of kilometres until the lower areas of gentler relief are reached in the centre of the continent.

The map shows how the pattern of rainfall varies over the Plains. Much of the weather is brought by air moving in from the Gulf of Mexico which lies to the south-east. As a result the rainfall tends to get less and less the further west one goes, with most of the rain in all these places falling in the summer. The diagram shows another important feature of the rainfall – its unreliability. Years of above-average falls may be followed by years of drought.

Because the land heats up and cools down more rapidly than the seas, the middle of large land masses usually have more extreme temperatures than coastal areas. North America is no exception. While summers in the middle of the continent are very hot, winters can be bitterly cold and swept by savage blizzards and heavy falls of snow. Drought, frosts, hailstorms and blizzards make the Great Plains an area of climatic extremes and hazards for any settlers.

The natural vegetation of places with this sort of landscape and climate is grass. The grass is longer and thicker in the wetter east and grows in short clumps in the drier west. The grasslands of the rainier east have been ploughed up and the land farmed with wheat, maize and cotton for many decades now. But the drier western Plains have been a problem area. Some Indian tribes riding the horses that had been introduced by the Europeans used to hunt the herds of buffalo that grazed the grasslands. They were followed to the Plains first by ranchers who grazed their cattle on open, unfenced ranges and then by farmers who tried to grow crops as they had done in the wetter east. The stories of the clashes between Indians and cowboys and ranchers and farmers are very well known. What is not so widely known are the disasters to ranchers, farmers and land resulting from overgrazing and crop growing in unsuitable places. All was well if the rainfall was high, but in years of drought the grass and crops failed to grow and the exposed soil was blown from vast areas, turning grassland and fields into desert.

Nowadays, by using irrigation, carefully selected drought- and frost-resistant seeds, dry-farming methods and controlled grazing, farming on the Plains is more secure than it was. But it still is an area of risk for the farmer, where farming may succeed or fail, depending on the chance of the weather.

1. The further west you go across the Great Plains the drier it gets – yet the California coastlands get plenty of rain. Why is this?
2. In how many of the years between 1915 and 1945 did Oklahoma have above average rainfall, and in how many did it have below average rainfall?
3. Choose either the Red Indian or the Homestead scene and list some of the clues it tells us about a) the Plains environment, b) the lives of these early migrants to the Plains.
4. 'Drought, frost, hailstorm, blizzard'. Describe how these could each be a hazard to the crop farmer.
5. The horse, barbed wire, the combine havester and the wind-pump have all had a big effect on the use and appearance of the Plains. Write a sentence or two on each of these saying what their effect was.

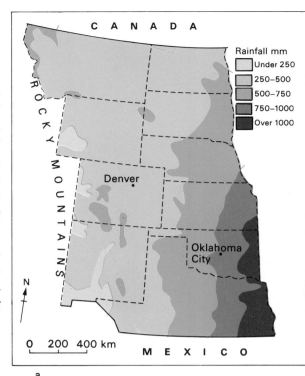

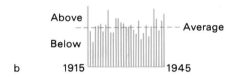

a) How the annual amount of rainfall changes from the Rockies to the middle of America. b) How the annual rainfall varied over 30 years in Oklahoma

Rounding up cattle in Texas   Wheat fields on the Great Plains

6 Mid-latitudes

Ocean margin and continental interior

Winter weather like this is typical in the Mid-West, on the same latitude as Greece or Southern Italy

There are no clear boundaries to the mid-latitude lands, but in general terms they lie between the warmer tropics and the cooler lands near the Arctic and Antarctic Circles. The areas between the latitudes 30° and 60° north and south of the equator give a rough guide.

The fact that many places within these latitudes have a fairly moderate climate – not too hot, nor too cold, not extremely wet, but not too dry – led them to be sometimes called the temperate latitudes. It is easy to see why people use that name on a typical spring or autumn day in, say, some of the remaining woodlands of Britain. It is also easy to understand why these lands were some of the first to be cleared and settled and turned into rich farmland. It was in these latitudes too that many of the world's cities first developed. But the label is dangerous because it gives too simple an impression of the wide variety of environments to be found.

The western half of the United States that has been described in the past few pages lies within the temperate zone. The many variations in environment have also been described. They range from 'Mediterranean' valleys, to harsh deserts and continental interiors with hot summers, very cold winters and a natural cover of grassland. From the map we can see that much of Canada that lies to the north of mainland USA, and much of Europe that lies north of the Mediterranean is also within the mid-latitude zone. Although it is everywhere a

A comparison between coastal and continental climates in Europe and the USA

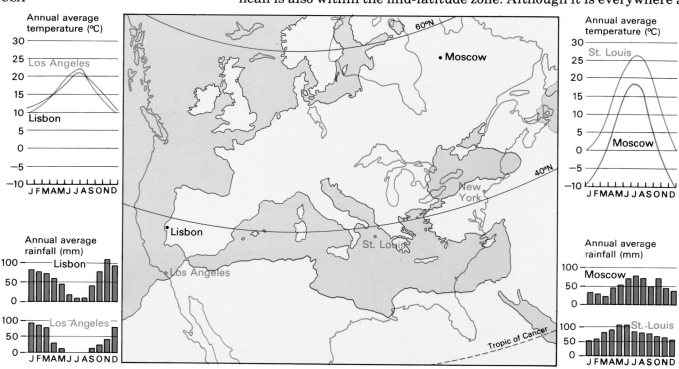

Downtown Manhattan, the Central Business District of New York

Below: A 'heat map' of the North-East coast of the USA. The Gulf Stream can be seen clearly in the bottom right-hand corner, coloured green. The 'hot' cities are pale grey

Bottom: The Eastern seaboard 'megalopolis'

little cooler than further south, the same differences between the ocean margin and the continental interior can be seen. The climate of the British Isles is much rainier and milder than that of eastern Europe and European Russia. In the same way that the forests of the Pacific coastlands contrast with the grasslands of the Plains and Prairies in North America, so do the mixed forests of western Europe contrast with the grasslands of the Steppes in the USSR.

It is not just nearness to the sea that makes a coastal zone mild. The temperature of the seas that wash the shores matters a great deal, and the British Isles are much more affected by the warm waters of the Gulf Stream (that can be seen in the 'heat map') than is the east coast of the USA. As a result, winters in Britain are milder than those of New York although the British Isles are much farther north. Even so, the deep inlets formed along the mid-Atlantic coast when the sea level rose after the melting of the ice sheets has provided magnificent sites for a line of huge cities. This is one of the most densely populated areas in the world, with the cities of New York, Philadelphia, Baltimore and Washington almost merging to make one giant built-up, urban environment 'megalopolis'.

1. New York, on the western side of the Atlantic, is only a few degrees further north than Lisbon which is on the eastern side of the ocean. New York's average temperatures for the warmest and coldest months are 23°C and 0°C. From the graph, say what they are for Lisbon.
2. Compare the temperatures and rainfall of Lisbon and Los Angeles with those of St Louis and Moscow. Give the title 'ocean margin' to one pair and 'continental interior' to the other.
3. The picture opposite is of farmland in the northern part of the central United States. Imagine you are the person walking along the track. Describe the scene and the weather.
4. Draw in the coastline of the north-eastern United States shown on the map to the right. Shade in and label the 'North Atlantic Drift'. Mark in and name any urban area you can see on the 'heat map'.

7 Mountain Land of the Incas

Running down the whole length of the western side of the continent of South America is the huge mountain range known as the Andes. Here are found some of the most spectacular mountain peaks, high bleak plateaux, lakes and steep-sided river gorges in the world. People have lived in the mountains and valleys of the Andes for thousands of years. One of the best known of these Andean peoples were the Incas, who lived in the area now known as Peru.

The Incas once controlled a vast Empire stretching northwards to the coast of present-day Ecuador, westwards to the narrow, dry Pacific coast, eastwards to the rainforested Amazon lowlands and southwards as far as the Atacama desert in modern-day Chile. All this was ruled by the Emperor and his royal family from their capital of Cuzco in one of the mountain valleys. In 1532 this empire was conquered by Pizarro, a Spanish adventurer, and a handful of soldiers.

One of the many settlements the Incas built was called Macchu Picchu. It was probably a sacred place for the worship of the Sun God, the god of the Incas. Whatever its purpose it certainly has a remarkable site, towering some 600 metres above the Urumbamba River, a tributary of the mighty Amazon. It was never discovered by the Spanish invaders, and was only spotted beneath its cover of dense vegetation by an explorer in 1918.

All the wealth of the Empire belonged to the Royal family and the state – and enormous wealth was available in the form of gold, silver and precious stones. A person's rights and standards of living depended very much on his or her place in the strict class system.

The Inca city of Macchu Picchu, high in the Andes

The site of Macchu Picchu

Inca craftmanship. Stonework built without mortar (*above*) and a ceremonial knife (*right*)

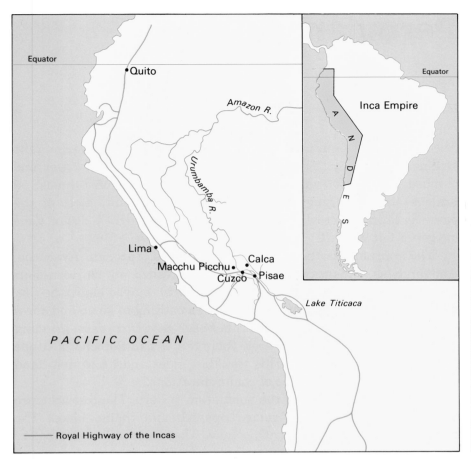

Left: The extent of the Inca Empire

The deep valley of the Urumbamba River

An Inca road

Most people lived in villages or towns scattered through the mountains. They grew maize and potatoes, or kept llamas and alpacas. These animals provided the mountain people with food and raw materials for clothing, as well as a means of transport.

The wheel was unknown, and transport was on foot or by animal. This makes it a bit surprising that the Empire was linked by an excellent road network. In flat areas the royal road would be about six metres wide, but in places it had to be cut into the sides of mountains, often as steps. Deep valleys and rivers were crossed by different sorts of bridges, including the frightening, steeply looped and swaying, rope suspension bridge.

The Incas were without a written language, machinery, the wheel, and many kinds of weapons, yet they planned and built magnificent towns and irrigation systems. Many of the stone buildings were so skillfully constructed that they are standing today.

1 Make a sketch of the scene of the Urumbamba river and label to show: narrow river floor; steep-sided gorge; mountain peak with precipitous sides.
2 Note the latitude of these Andean areas of Peru. What sort of environment lies to the east? Why is the vegetation cover different here? Give at least two reasons.
3 What is good and what bad about the site of Macchu Picchu as **a**) a fortress, **b**) a centre of trade for the surrounding area?
4 What are some of the difficulties of life in a mountainous area like the Andes?

7 Mountain Peruvian Andes

A Quechua Indian with his herd of Llamas

Below: This is the sort of rugged scenery in which the journey described on this page took place

The following account is of a journey made from Lima to a farm in the Urumbamba valley: 'The train to Huancayo left Lima at 7 a.m. and climbed up to 4 800 metres in 8 hours. There are 66 tunnels, 59 bridges and 22 zigzags. Towards the highest point, most of the passengers in the crowded carriages had passed out. A doctor gave oxygen. I felt very ill.

'The journey to the farm took ten days, through Ayacucho, Andahuaylas, Abancay and Cuzco. I travelled in thirteen-hour stretches, in buses, standing up in trucks with maybe eighty people, and the last stretch in a hired taxi. It was nothing to go up and down several thousands of feet several times each day. Dirt roads, grinding gears and the smell finally took their toll. I had Soroche, altitude sickness, for several days. This was the Andes and I had first-hand experience of the difficulties of communications.'

Indians form about half the population of Peru. The rest are pure Spaniards, or Mestizos of mixed Spanish and Indian blood. The mountain Indians are mostly Quechuas. They were in the area before the Incas and Spaniards conquered them, and have a special physique to cope with the environment. They have smallish bodies, large chests and lungs, and an adapted blood structure to cope with the thin mountain air.

Most Quechuas lead an extremely hard life as village farmers. They grow crops and keep llamas and trade as best they can in local markets. Chewing coca is universal. The leaf contains cocaine and acts like an anaesthetic to the stomach. It enables the Indians to work for a long time without food and resist the cold. In some places a little extra money is made by selling it illegally. The small towns are not as clean as most in Britain. They may have electricity, shops, a church,

Market day in a small Peruvian town

A sign of change. This separator and smelter for various minerals is found at La Oroya

police station, school and so on, but frequently they are without running water or sewage facilities – the main stream is used for bathing, washing and drinking.

But some changes are taking place. In the Urumbamba Valley, for example, the old Spanish 'haciendas' on which the Indians worked for very little are beginning to be replaced by co-operatives and freehold farms. The Indian villages are sometimes able to buy a lorry to take their produce to market. Cuzco itself is quite a modern town and tourist centre. There are also a number of large mines and smelters owned either by foreign companies or wealthy families in Peru that provide a different sort of work for the Indians. But whether as peasant farmers, farm labourers, workers in mines or in factories, life in the mountains is still extremely hard for most Indians and their families.

1. From the map draw the route of the journey from Lima, through La Oroya and Huanuco to Pucalipa. By the side of each place name, add its altitude.
2. What effect did travelling in the mountains have on the English visitor? Why does it not have the same effect on the Quechua Indians?
3. Measure the distance from Huancayo, the end of the railway journey from Lima, to Cuzco. It took about eight days to do this journey on the mountain roads, travelling in thirteen hour stretches. How far could you travel by car at an average speed of 100 kilometres per hour in that time?
4. Imagine you were the young Indian looking after the llamas in the photograph. What would be the advantages and disadvantages of staying in the village as a farmer or going to work as a miner or worker in the ore plant? What other possibilities might there be for your future?

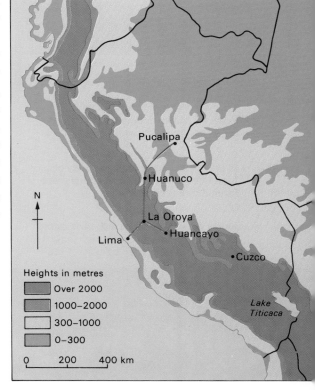

The Peruvian Andes, showing the route from Lima to Pucalipa

A cross-section of the route across the Andes from Lima to Pucalipa

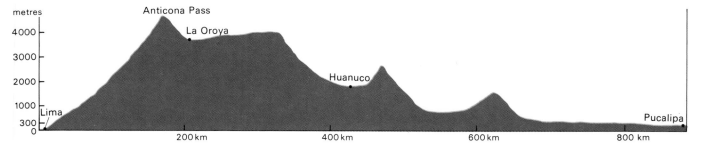

7 Mountain Building up and wearing down

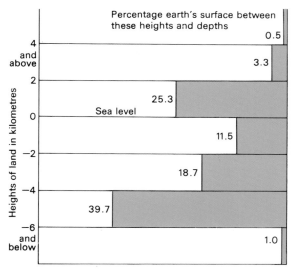

The percentage distribution of the earth's surface, in heights above and below sea level (shown in green)

Below right: The U-shaped valley of Lauterbrunnen, a typical Alpine scene formed by the work of a glacier

Below: The Swiss Alps: a good example of 'young' mountains with high jagged peaks and steep valley sides

On a true-to-scale model of the earth the great mountain ranges would show as tiny bumps and ridges. Yet as the photographs of the Himalayas (page 13), the Andes (pages 56 and 58) and the Alps (page 114) show, mountain environments can be spectacular. They are often so rugged and so high that it is impossible for large numbers of people to live permanently in them.

The diagram on this page shows the percentage of the earth's surface that is land and water, and the amount of this surface that lies between certain heights and depths. The one opposite shows the latitude of the highest mountain peaks, while the map shows the distribution of the world's main mountain ranges.

The way in which the earth's crust consists of a number of massive 'plates' was described on page 10. It seems that at certain times in the past the movements of these plates forced great thicknesses of sedimentary rocks that had been laid down in seas into huge folds. These folded, crumpled and broken masses of rock were raised many kilometres above the level of the sea. Because of the way they were formed they are called fold mountains. As soon as they were formed, the atmosphere, wind, rain and ice began to wear and carve them down. Enormous masses of material from the mountains have fallen, been blown or been carried by rivers and streams into the lowlands and finally the seas and oceans – to make new sedimentary rocks. The oldest fold mountains have been worn down the most, and on the whole it is the younger fold mountains that are the highest and

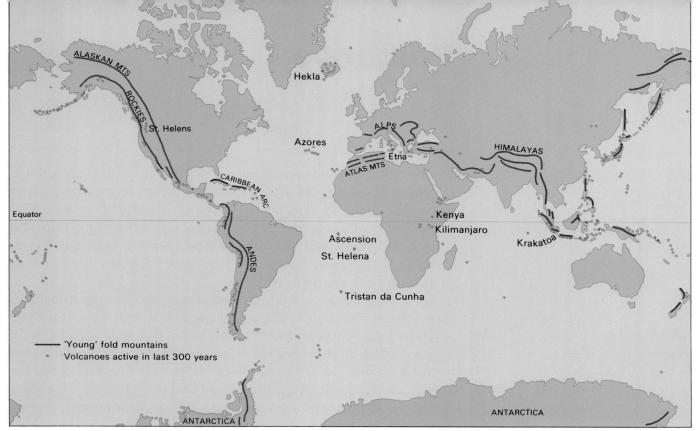

provide the most striking landscapes of jagged peaks, steep slopes, ridges and valleys filled with glaciers. It is as hard to believe that mountains like these will one day be worn down to gentle landscapes as it is to believe that they are made of rocks that once lay at the bottom of deep oceans! It is a little easier to understand if we realise that the 'younger' fold mountains were being formed about forty million years ago! It is also important to remember that many mountains of the world have not been formed this way.

An obvious disadvantage of mountain environments to man is the ruggedness of the slopes. Another one is not obvious to see from photographs, but very easy to recognise in real life. The air gets thinner with height. This not only makes breathing more difficult – oxygen is vital above certain heights – but means that the surface heats up and cools down more rapidly than at sea level. Mountains are also often the setting for violent winds and storms, as well as heavy snowfalls.

1. Which continent has the highest mountain range a) in the northern hemisphere, b) in the southern hemisphere? Where are the highest peaks on the equator?
2. What do you notice about the height at which trees stop growing as you move from the Poles towards the equator? Explain why some of the mountains at the equator can be snow-covered.
3. Make a sketch of the alpine scene and label with the words: frost-shattered peak; knife-edge ridge; steep cliff face; glacier-river of ice.
4. People have very different feelings about mountains. Write a few sentences about any of the mountain environments you know, or have visited, or that are shown in this book. Try to describe your feelings rather than just the scenery.

The world distribution of major active volcanoes

The heights of mountain ranges at different latitudes

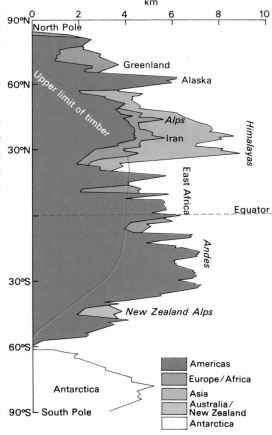

7 Mountain **Volcanoes**

Mount Etna in Sicily, one of Europe's most active volcanoes

A close-up of the almost perfect cone of Mount Mayon in the Philippines

The beautiful cone-shaped mountain in the Philippines is not a fold mountain. It was formed in a completely different way. Because of the great heat, rock at a distance below the earth's crust is molten. This molten rock, called magma, sometimes finds its way to the surface and is forced out through cracks or vents. The magma is then called lava. The cone-shaped mountain is a volcano, made from lava that has come from beneath the earth's crust.

Lava varies a great deal in composition and the way it looks. If lava is thick and pasty, the trapped gas mixed with it explodes out, shattering the lava to fragments. The red-hot molten lava pours out of the vent while large fragments of lava, ash and fine-grained dust are hurled high into the sky to fall back on top of the vent. Because the lava is pasty it forms short, thick flows that pile up around the vent. The lava fragments and ash from the explosions collect round the vent or crater in beds that get thinner away from the centre. So steep-sided volcanic cones such as the ones in the picture are formed with slopes up to about 30° leading up to the main peak. If the lava is runny it cannot form a cone shape but spreads like sheets over the land or ocean floor.

Volcanoes are normally active for short periods, perhaps a few months or a year at a time. Between eruptions they remain dormant or sleeping for maybe several hundred years. During this time the

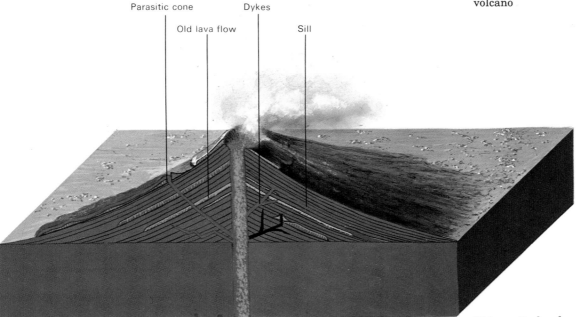

Block diagram of the cone of an active volcano

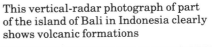

This vertical-radar photograph of part of the island of Bali in Indonesia clearly shows volcanic formations

All that is left of this volcano is the hard central plug. Le Puy in France

lava in the crater solidifies, plugging the vent or crater. Gas pressure builds up until there is an explosion that bursts open the plug and the lava flows through the vent again. A volcano is said to be extinct when it has not erupted in historic times – though some that were thought extinct have surprisingly erupted with disastrous results.

At first sight it seems odd that people should want to live near volcanoes that might erupt, but a glance at the photograph shows that rich crops can often be grown on the lower slopes of volcanoes. The climate has to be suitable, of course, but when that is so the volcanic lava and ash usually break down under the weather to provide very rich soils. So in many parts of the world within the tropics or in mid-latitudes there are many people farming the lower slopes of extinct and sometimes even active volcanoes.

Like any mountain that is exposed to the weather, the surfaces get eroded by rain, wind and ice and broken up by temperature changes. In time only the remnants of the volcano may remain – perhaps the central plug is all that is left.

1. Draw a sketch of the volcanic cone in the Philippines. Label to show the crater, erosion of the slopes, cultivated land on the lower slopes and surrounding land. Give a title, including the latitude and longitude of Mount Mayon on Luzon Island (see map on page 77).
2. Look at the map showing the distribution of volcanoes active in the past few centuries on the previous page. Compare it with the map showing the main plates and their margins on page 11. Can you see any link between the distribution of these volcanoes and fold mountains? Can you see any link between the distribution of the volcanoes and mid-ocean ridges? Give the name of one mountain range or line of volcanoes in each case.
3. Suggest two very different reasons why the upper slopes of extinct volcanoes are not suitable for settlement and crop growing.

8 Water resources

Water: the basic resource

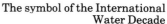

The symbol of the International Water Decade

A traditional method of irrigation – Luxor, Egypt

For many children around the world, whether in Africa (*top*) South America (*middle*) or India, water is a scarce resource. Sometimes they have to walk long distances to get water and, even if there is a tap, the water may be polluted

The amount of water needed by human beings to survive is surprisingly small – about two litres, or four pints, a day, although this varies with the climate. But as well as having the right amount, the water needs to be clean and free from disease-bearing bacteria. Many diseases are carried in untreated water. It often comes as a surprise to people living in Britain, who usually have only to turn a tap to get as much clean water as they want, that millions of people around the world cannot be sure of a shared supply of regular, clean water – let alone a supply of their own.

Besides these minimum bodily needs, though, water is used for many other purposes. In Britain domestic consumption of water began to increase rapidly in about 1850 with the growth of sewage systems and piped household supplies. In all developed countries, the growing use of water-using household equipment has sent up consumption even more in recent years – this includes such 'luxury' items as lawn sprinklers and swimming pools! The domestic water consumption in the USA is about 450 litres a day per person. In Britain it is lower at about 200 litres, although this is expected to rise to about 300 litres by the end of the century.

The ways in which water is used for irrigating farmland in different countries has already been described on page 48. The methods used may range from simply building mud walls around fields to complicated furrow or sprinkler systems. The source of water can be rainfall, underground supplies brought to the surface from wells, or surface water flowing in rivers or trapped in huge reservoirs by giant dams. A vast amount of water is also used in industry.

Sometimes this is to generate power in massive hydro-electric power stations, sometimes for cooling and sometimes in the manufacture of certain products, such as paper.

Man has known for a long time how to trap rainfall or water from flooded rivers, and recently has been able to build great reservoirs to hold the water until it is needed. But as the population of the world grows, more water is needed and new methods of obtaining and storing and reclaiming water are being tried. In some desert areas factories have been built to turn salt sea water into fresh water, but the costs of running these desalination plants is high. In dry areas clouds have been seeded with crystals of silver iodide to produce rain – but this can't work in cloudless skies! There are even plans to use the water locked up in the ice sheets and glaciers, but this has yet to be tried on a large scale.

For a long time to come man will have to rely on rainfall and existing river flow and work out better ways of storing and reusing the water – hence the need to avoid polluting rivers as much as possible. There is also the desperate need to provide millions of people with a supply of fresh water to make their lives better and safer.

A desalination plant in Saudi Arabia. You can see how complicated the machinery is for converting sea water into drinking water

1. List the different ways you and your family used water during the past week. Say where the water supply comes from, who provides it and who pays for it.
2. Look at the three photographs on the far left. In each case say what are the problems or dangers the young people have because of the lack of an adequate supply of water.
3. Look through the book and give examples of water being used to irrigate land, mentioning where the countries are. Six countries account for over 70 per cent of all the irrigated land in the world – China, USSR, India, USA, Pakistan and Iran. In which continents are these six countries?

Cairo, capital of Egypt on the fertile River Nile. Cities use vast amounts of water

8 Water resources

The River Nile

The Blue Nile in its early stages in the Ethiopian Highlands

The River Atbara in flood in September. At other times of the year scarcely any water will be flowing

A felucca, the traditional Egyptian sailing boat, on the Nile

About ninety-seven per cent of the world's water is contained in the oceans, but since it is salty it cannot be used for drinking or irrigation without costly treatment. Almost three-quarters of the remaining three per cent is the fresh water in the ice sheets of Antarctica, Greenland and the Arctic Ocean and in smaller ice sheets and glaciers in mountains. It is on the remaining amount in rivers, lakes and in the ground that people depend for most of their water needs.

The River Nile is one of the longest rivers in the world, but it has a very small average discharge and its waters do not even reach the sea during three months of the year! Both the White and the Blue Niles rise in tropical latitudes where rainfall reaches 2 000 millimetres a year, but after they join at Khartoum to form the Main Nile the river flows through one of the largest and driest deserts in the world. North of the River Atbara no further tributary enters the Nile during the remainder of its 2 000 kilometre course to the sea.

The tributaries coming from the south and west have gentle slopes and flow through either wooded country, or great lakes like Victoria or swamps like the Sudd. Although it does flood its banks, the White Nile doesn't vary in flow all that much throughout the year, while the gentle slope means only the finest silt gets carried in the river as far as Khartoum. The Blue Nile and eastern tributaries are very different. These rivers rise in the mountains of Ethiopia through which they flow in massive and deep gorges. Their flow varies from a roaring, brown-coloured flood up to 800 metres wide in late summer and autumn to a mere trickle in spring and early summer. During their flood period they carry masses of coarse silt into the White Nile.

Seasonal rates of flow on the River Nile

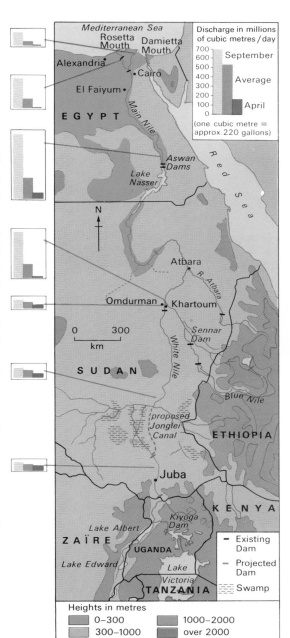

The silt in the Nile has built up the magnificent delta on the shores of the Mediterranean seen in the photograph. In the past, when winter discharge was low, earth barrages had to be built across the distributary channels to stop the salt water entering the delta. A permanent barrage has now been built at Edfina with gates designed to keep river and sea water apart until the river flow is enough to do so on its own. Settled agriculture and town life began in the Nile delta several thousand years ago, and the annual floods that brought water and fresh silt to the river banks enabled the desert to be farmed in a narrow strip along the Nile almost to the Sudan border. Even in the rainier lands of the southern Sudan, where cattle are kept by nomadic pastoralists, the people are dependent on the White Nile floods for water and grazing during the dry season. During this century a great deal has been done to harness the waters of the Nile for the good of all the people along its banks. But doing so raises the question of who owns the Nile, and who should decide what happens to it and who should benefit from any developments.

1. Look at the diagrams of river discharge on the map. a) Compare the discharges of the White Nile near Juba and the Blue Nile near Khartoum. b) Compare the average flow just below the Aswan Dam with the combined average flows in the delta – what has happened to the difference in water amounts?
2. Put into words what you see and feel about the three scenes of the Nile and its tributary in the photographs.
3. Make a sketch of the air view of the delta. Label it to show the desert, irrigated farmland, the Mediterranean Sea, the Rosetta mouth and the Damietta mouth, the location of Alexandria. Give some idea of the size of the delta on your sketch.

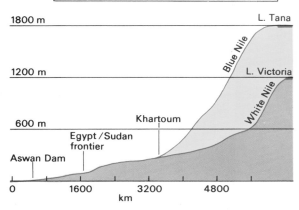

Profiles along the Nile

Left: A satellite view of the fertile Nile Delta. The contrast between the cultivated areas and the desert is marked

8 Water resources — Using the Nile waters

The Aswan Dam

An Egyptian village in a quiet sidewater, running off the Nile

Crops need water and food from the soil if they are to grow and produce high yields. Irrigation water and a fresh layer of silt, rich in plant foods, was provided each year by the Nile floods for centuries. Thousands of villages were built in the delta and on low hillocks along the river banks, and every scrap of land used to grow crops of rice, sugar cane, cotton and vegetables. Nile water was used for domestic purposes while thousands of feluccas sailed on the river carrying goods from place to place. Egypt depended on the Nile.

One problem was that the annual floods were not reliable, and as the population grew there wasn't enough water for everyone. So a number of dams were built. One was built at Aswan in Egypt, another at Sennar in the Sudan and a third at the Owen Falls in Uganda. But still these were not enough, and in the 1960s Egypt, with help from the USSR, built another larger dam at Aswan. The huge Lake Nasser formed by the dam provides a vast reservoir of water for irrigation, for power and for use in the cities. But apart from its cost the project has produced some difficulties and problems. In the heat and arid climate of those latitudes there is a great loss of water from the lake by evaporation. The silt that once flowed over the farmland is now deposited in the upper part of the lake, not only making it smaller, but also making it essential to use fertilizers on the farmland.

The countries of the River Nile

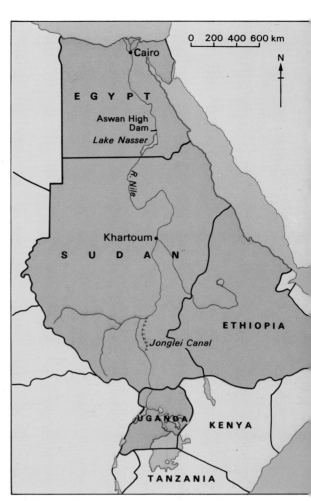

Both Egypt and the Sudan continue to need more water as their populations grow. It would seem easy enough to release water from Lake Victoria through the dam at Jinja – but that would need the agreement of Uganda, Kenya, and Tanzania, who all have shores around the lake. The water from this lake then has to flow through the Sudd, the swamps of southern Sudan, and about half the flow of the Nile is lost in the process. It is to try and provide a more direct and less wasteful flow that a big canal – the Jonglei Cut – is being made in Southern Sudan. Egypt and Sudan have agreed to share the extra water available, while another benefit would be that the now-flooded lands of the Sudd could be cultivated and irrigated. The people of the southern Sudan, such as the Dinka, are cattle farmers, however, and may not wish to change their way of life. There is also a danger that if there was a year of heavy rainfall the improved flow of the canal and river would lead to widespread flooding in the centre and north of the Sudan.

The control of the Nile, then, is complicated because the river and its tributaries flow through several countries, and what one country does affects the others. There are also very high costs, technical problems of flood control and social problems involved in changing the way of life of many people.

1. If it were not for the Nile, the area shown in the photograph on the left would be desert. Make a labelled sketch of the scene to show how it is affected by the nearby Nile and its waters.
2. Draw a line five centimetres long, representing the 500 kilometres or so between London and Edinburgh. Alongside it draw Lake Nasser and The Jonglei Cut at the same scale.
3. Would it be right or wrong, wise or foolish, for either Ethiopia or Uganda to try to control the headwaters of the Blue or White Niles so that Sudan and Egypt could not use the water? Explain your answer.

Below: The huge bucket wheel used for cutting the canal through the Sudd

The Sudd of the Southern Sudan consists of thousands of square kilometres of swamp in which there are numerous 'islands' with small villages like this

8 Water resources

Using the rivers of northern Siberia

The hydroelectric power station at Krasnoyarsk on the Yenisei

The Yenisei is one of Siberia's main rivers

A river in North-Eastern Siberia

The USSR has the largest fresh water resources in the world. With the exception of the Volga however its largest rivers flow across the sparsely inhabited and less developed northern and eastern lands which use little water. The most populous and economically important parts of the USSR are in European Russia, Kazakhstan and Central Asia. Water use is very heavy in these areas but the river discharge is less than twenty per cent of the national total. There is a lack of balance between demand and supply, so the government is planning to move water from places where it is not needed to where there is an increasing shortage.

One of the most ambitious schemes is to take water from the River Ob, one of the great rivers that cross Siberia, and its biggest tributary, the River Irtysh, and move it southwards into the arid lands of Kazakhstan and Central Asia. Additional water would be pumped from the River Yenisei into the Ob. The water would be channelled southwards through the natural depression called the Turgay Gate and into the Aral and Caspian Seas.

The first plan required the building of large reservoirs covering thousands of square kilometres. This plan was dropped when it was realised this would damage large areas of coniferous forest and its timber, flood valuable farm land further south and increase waterlogging in the already swampy region. Instead a plan was drawn up by which a number of dams and power stations would lift the water into the huge Ob-Caspian Canal. When completed this will stretch 2 300 kilometres to rivers flowing into the Aral Sea, and probably on to the Caspian Sea. The canal will be between 120 to 170 metres wide

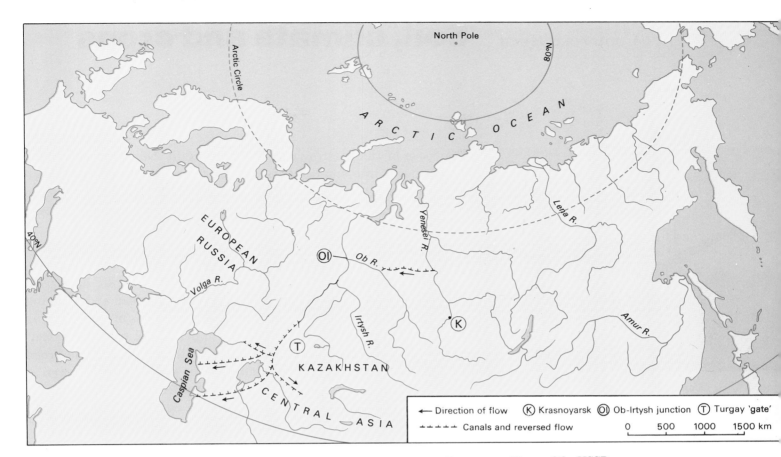

Rivers of the USSR

and average 12 metres deep – needing the removal of several billion cubic metres of earth and rock.

The scheme will in effect reverse the flow of the River Irtysh, and water will be pumped into this from first the Ob and, when the project develops further, from the Yenisei. The aim of this scheme – unlike that at Krasnoyarsk further upstream on the Yenisei – is not to make power. In fact vast quantities of electrical power will be needed to drive the pumps that will raise and move the water against its natural flow. There is also the possible damage to the environment, not only the forests and tundra, but also the Arctic Ocean itself. If the Arctic Ocean pack ice was greatly affected it could alter the climate of the northern hemisphere. Because of the enormous scale of this and similar projects, it will be decades before they start. Sooner or later, though, because water is such a vital resource not only for human survival but for modern industry and city life, the projects will be completed.

1. Make a drawing of the Krasnoyarsk power station and label to show: the River Yenisei; the dam; the coniferous forest on the hills in the background; the pylons and cables carrying the electricity to the industrial towns; water being allowed through one of the slipways near the far bank.
2. Compare the latitudes and climates of the mouths of the Rivers Nile and Lena. What will be the different problems of river flow in the mouths of the two rivers at the most difficult times of the year?
3. Why is it easier to make a plan for the control of the Ob and Irtysh than for the Nile?

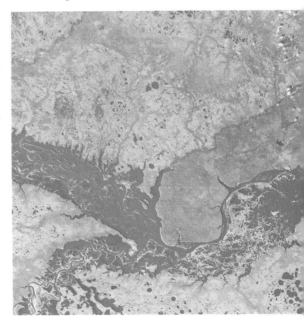

A 'Landsat' photograph of the Ob and Irtysh Rivers. In such photographs, features on the ground (such as vegetation) are picked out in special colours. The blue areas are liable to flooding

9 Crop resources

Soil, climate and crops

Combine harvesting wheat. Huge machines are used in this style of 'extensive' farming

The rich, black soil lands of the Prairies

The rocks of the earth's crust are continually being 'weathered' or broken down into small stones and tiny particles. While this loose material may stay where it was formed, on steep slopes it is more likely to fall, slide or roll downhill. On lower land with less steep slopes it is still likely to be moved around by wind, water or ice. In whatever way it was formed, rock debris of one sort or another provided the soil particles and mineral content of the world's soils.

The spaces between this mineral mass of particles, are filled with air and water and are usually also full of living or dead and decaying plant and animal matter. This once-living material is broken down by small insects and bacteria. It is this mineral and organic matter, dissolved in water, that provides plants with their food. The green plants make carbohydrates by combining the hydrogen and oxygen in the water with carbon from the atmosphere. Human beings then eat these plants and their bodies convert the carbohydrate back into energy.

The size of soil particles varies from the large ones found in sands and gravels to the tiny ones found in clays. Sandy soils easily lose their soluble minerals as the water drains down through them. Clays hold the water and plant food more easily – but the soils are quickly waterlogged. Soils with textures in between sands and clays are called loams, and they have the advantages of both the others.

It has long been realised that it is no good planting seeds in the soil and hoping for the best! Men have learned how soils and climates differ, and how each plant needs certain conditions to grow well. They have learned how to prepare soils, manage them before and during

Green plants turn carbon, hydrogen and oxygen into carbohydrates, vital for animal growth, by using energy from sunlight

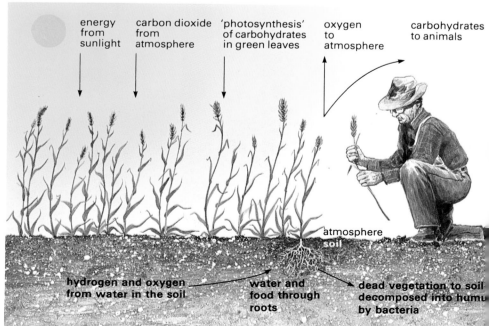

A wheatfield in the American prairies

plant growth, and how to replace plant foods that have been used up by the crops. They have also learned how to make up for some of the shortcomings in the climate and weather, to protect the growing crops from disease and pests and even to breed new varieties that give higher yields than the original ones.

There are many different climate-soil-vegetation patterns around the world that have evolved over thousands of years. The mid-latitude climates and rocks of North America and the USSR, for example, have produced a grassland vegetation on a rich, black soil, known by its Russian name of Czernozem. The upper parts of this soil are rich in plant foods, which is why tall grasses grow so well. These particular soils and climates have been used for a long time now to grow wheat on a large scale – wheat is a variety of grass that provides an excellent food for man. At first the same crop was grown year after year, a practice called monoculture. This reduced the richness of the soil. So now farmers alternate wheat growing with other crops that return some of the plant food through their roots. They also add chemical fertilizers. In these ways they manage the soils to provide essential crop resources.

Czernozems, the soils of the Prairies of North America and the Steppes of the USSR

1 Wheat is a very important and basic food. List some of the ways in which the wheat grain is processed and treated before we eat it.
2 Growing crops over a large area with few people and machinery is known as 'extensive' farming. Growing crops in a very small area, usually with a large number of people and using every scrap of land is known as 'intensive' farming. Look back through this book and give two examples of each. Give one advantage and one disadvantage of each sort.
3 Wheat is a cereal crop that can grow in quite a wide range of climates, but the really important areas are in the mid-latitude grasslands. What does this tell us about the sorts of climate the wheat plant doesn't like? In the northern parts of the Prairies the wheat is called 'spring' and further south it is called 'winter'. What does this tell us about when the two varieties are planted? Why the difference?

| 9 | Crop resources |

New land and controlled climates

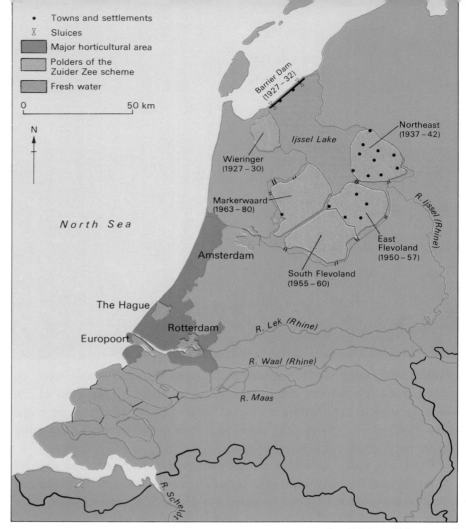

The Dutch polderlands

The Netherlands lies where three great rivers, the Rhine, Maas and Scheldt form a large delta before entering the sea. For over a thousand years the Dutch people have built barriers or dykes to keep the sea and rivers from flooding the low-lying land. First windmills then steam, electrical and diesel pumps have been used to drain the flood waters from this land and lead water by canal into rivers and the sea. Today about half of the Netherlands consists of reclaimed land, known as polders. Much of this is below sea level.

The biggest and most recent of these schemes, started in the late 1920s and completed in the early 1980s, was the reclamation of a large part of the sea inlet known as the Zuider Zee. The first task was to turn the sea inlet into a freshwater lake (Ijssel Lake) by building a dam across the mouth of the Zuider Zee. The enclosing dam is over thirty kilometres long, and is wide enough to carry a road and a railway if necessary. The five polders created out of the bed of Ijssel Lake increased the area of cropland in the Netherlands by ten per cent.

These man-made polders have rich, deep soils and are used to grow many crops, including grass. With modern methods of drainage, the water table can be kept low enough for powerful farm machinery to be used. A typical farm on the polderlands grows wheat, barley, sugar beet, potatoes, and flax. The fertility of the soil is kept up by rotating these crops so that no one of them is planted in a field for many years without a change. A great deal of chemical fertilizer is also used.

Most of the coast between Europoort and Den Helder consists of tall and wide sand dunes. Behind these, especially in the area between Rotterdam and Amsterdam known as south Holland is a remarkable farm landscape. Every bit of land is carefully used to grow vegetables and soft fruits, flowers and bulbs. This is one of the most intensive forms of farming in the world, with much of the cultivation being done by spade and hoe. It is more like gardening than farming at times and is known as market gardening, or horticulture. The cost of land is high, and a lot of money is needed to provide glasshouses, heaters, humidifiers, and sprinklers. But the cash returns are high for these forms of farming. Temperature, light, atmospheric humidity and soil water are all carefully controlled in the artificial environment of the glasshouses. All this is very different from extensive grain farming on the Prairies!

One reason for this sort of farming is nearness to a market. Millions of people live in near-by Dutch and German cities. Another is the nature of the rich, light and easily worked soil provided by the mixture of sand, peat and clay. But there is also the skill and tradition built up over many centuries, as well as the money to provide all the equipment and scientific help needed in modern commercial horticulture and market gardening.

1. What are the main differences between the farmlands on the North Eastern Polder and the Prairies, shown on page 73? What are the similarities and differences between the intensive farming near the Hook of Holland and that in the Great Valley of California shown on page 48?
2. Draw a simple bar graph or divided circle to show the following land use of the North East Polder: Arable land 83 per cent; Grassland 17 per cent. Compare it with the diagram showing types of farming. How do you think the grassland is used?
3. Name some of the products from a) orchards, b) market gardens, c) greenhouses. What are some of the advantages of these sorts of farms being fairly near large centres of population – or at least with good transport links with them. Name some orchard and market garden products for which nearness to centres of population is less important than for other market garden and greenhouse produce.

Glasshouses stretch as far as the eye can see towards the Hook of Holland, near Europoort

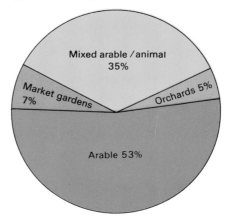

Types of farming on the Dutch polders

Left: On the reclaimed polderlands of Holland the farms are neatly laid out along the main roads

9 Crop resources

Rice: more cropland, richer yields

Rice growing requires flat land, so the mountain people of Luzon in the Philippines have built up terraces such as these over thousands of years

Rice is the main food crop for millions of people. Its ideal growing conditions are different from those of wheat. The most important variety, 'wet' rice or padi, needs to stand in fields covered with water during its growing season. The plants also need at least three months of fairly hot weather, with average temperatures over 20° Centigrade, and a dry sunny spell during the final stages of growth. As we can see from the climatic graph on page 35, the monsoon climate offers the best conditions, and about ninety-five per cent of the world's rice is grown in areas shown on the map.

Padi fields need to be fairly flat in order that water will remain on them for a sufficient time. The water must be prevented from draining away and a good supply must always be readily available. This is most easily arranged in river deltas or along river flood plains, and that is where most wet rice is grown. But in some places in mountainous countries these conditions are made by building elaborate terraces. These magnificent terraces in the Philippines have been developed over the past 2000 years. They are far older than the polders, and unlike them are still largely cultivated by hand or with the use of animals such as buffalo. The padi seeds are usually planted in special seed beds where they are carefully tended. When the monsoon rains fall bundles of young plants are transferred to main terraces as in the photograph. It is a hard back-breaking job, traditionally done by women. When the crop is ripe the land is

Transplanting rice seedlings by hand on the Philippine rice terraces

The 'water buffalo', as its name suggests, is particularly suited to work in rice fields

drained, the padi harvested and the grain threshed from the stalks, usually by hand. In some places several crops can be grown in a year. The poorer subsistence farmers will put some seed aside for the next planting and eat the rest – praying that there will be enough to last until the next harvest. The wealthier ones may have some spare to sell, while in certain areas large amounts are grown commercially to sell to people in the cities or perhaps to export to other countries.

In tropical countries staggering amounts of grain are lost after harvest from disease or through being eaten by insects, rats and birds. Improved storage could greatly increase the amount of grain available for food. Other attempts to do this have been by introducing new varieties of rice that give higher yields. A decade or so ago there was talk of the 'green revolution' and a hope this would solve the food shortage as populations grew. But it has not been completely successful, as the diagram shows. Although more rice may be grown, millions of farmers are little better off than before the 'green revolution'.

1. Look back to page 35. What is it about the monsoon climate that makes it suitable for rice growing?
2. Make a sketch of the scene of transplanting the rice into the terraces. Label to show: terrace, retaining mud wall, bundles of young seeds, water-covered soil, transplanted seedlings.
3. Both padi farming and market gardening on the polders of the Netherlands are intensive types of farming, but they are different in many other ways. Describe some of these differences, including the ways in which farmland has been made by man.
4. It is sometimes said that the Green Revolution is not a success because although food production for a country might rise, 'the rich get richer while the poor get poorer'. How could this happen? How might 'the poor' be helped to benefit from the developments?

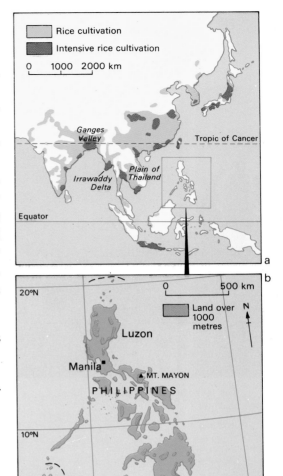

a Rice production in the Far East
b The Philippines: a rice-growing country

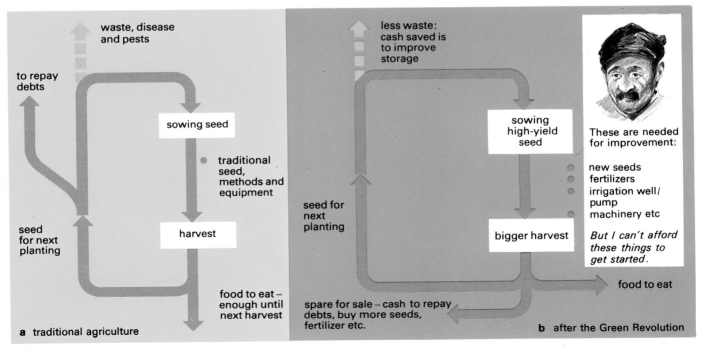

The nature and results of the 'Green Revolution'

9 Crop resources — Tea: a plantation crop

Tea pickers on a steep Sri Lankan hillside

Worldwide distribution of plantation crops and grapes

Many of our foods and raw materials are provided by bushes and trees. They are too important to be left to grow wild, and most are planted and cared for on estates or plantations. Tea is one of these crops, and Sri Lanka one of the most important tea growing areas in the world. The estates were made by British settlers or companies over a hundred years ago when Ceylon, as it was then called, was ruled from Britain. Apart from the carefully laid out and managed bushes there is a factory where the green, freshly-plucked leaves are processed and packed into tea chests to be sent to Colombo harbour, a network of roads and tracks linking the different parts of the estate and 'lines' of housing for the estate workers. The tea estate is like a combination of a large farm, a factory and a small village. But things have changed on the Sri Lanka tea estates, as this extract describes.

'Sri Lanka's tea industry needs a local labour force of over 200 000 skilled workers to replace the Indian Tamil immigrant workers who are being repatriated (sent back) to India. Despite several measures to improve working conditions on estates to attract good workers, not many Sinhalese have been willing to work on the plantations, partly because wages are lower than in other trades, and partly because it involves hard work under harsh climatic conditions. In the central highlands called 'Little England', where the world's finest teas grow at between 1 000 and 2 500 metres above sea level, it is bitterly cold for several months of the year.

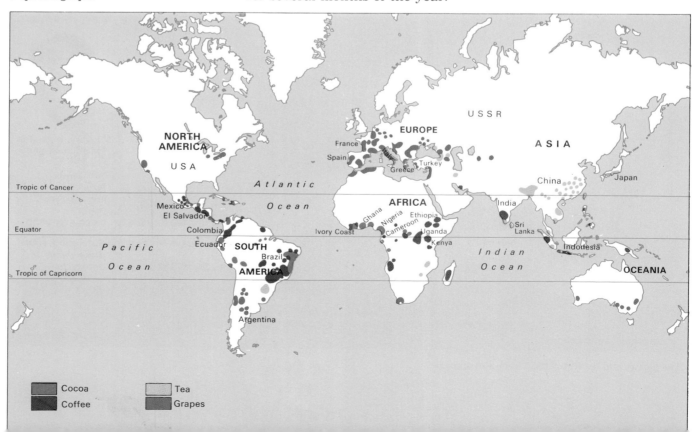

A close-up of Tamil tea pickers in Sri Lanka. Picking the right leaves at speed requires great skill

The 'lines' or rows of houses of Tamil tea pickers on a Sri Lankan tea plantation

In estates where Sinhalese have replaced Indians, both output and quality have fallen over the years, but the loss has been offset by the high prices tea is fetching on the world market, due to the sharp increase in the price of coffee. Many traditional buyers have warned against the drop in quality. Australia, which was one of the biggest buyers of Sri Lanka teas, has now switched to Indonesia. Sri Lanka's high-quality teas are bought largely by multinationals to blend with inferior teas from other countries.

Tea needs daily care. When plucking is delayed by even a few days, it affects the leaves and therefore the quality. The process of crushing and drying needs labour with long experience. An estate superintendent said that Sinhalese labour often failed to cope with the rigid time schedules for plucking and processing. This aside, they did not accept the regimentation and discipline which had contributed to the efficient management of estates under private ownership until their nationalisation in 1974.'

1. Look at the map and photographs, and read the extract. What clues are you given about the conditions needed for tea to grow well?
2. Draw a diagram to show the following information about Sri Lanka's income from the sale of goods to other countries – tea 50 per cent; rubber 20 per cent; coconuts and other agricultural products 15 per cent; non-farm products 15 per cent. Why are these figures to be used with caution? Why are they not likely to be absolutely true 'facts'?
3. What is meant by a) the tea 'industry', b) immigrant workers, c) nationalisation of the tea estates and d) repatriation of Indian Tamils?
4. The wealth of the country – however it is shared out – depends a great deal on the sale of tea to other countries. From the extract list some of the things that might a) increase, b) decrease the income from the sale of tea abroad. Give two examples of each.

Tea sorters in a Sri Lanka factory

10 Animal resources
Friends, and food

These early rock paintings from the Sahara show that hunting has been an important activity for man for many thousands of years

A reindeer herd in Lapland

Man has always hunted or used animals for some purpose or other. In many places, without the resources they provided, human beings would not have been able to survive. Man's basic needs are for water, food, clothing and shelter, and with the exception of the first of these – water – animals have been great providers.

At first animals were hunted as they roamed wild and free, and man had to develop great skills as a hunter to obtain the food he needed.

Very few people need to hunt in order to survive these days but there are still quite a few people who herd animals. They move them from place to place in search of pasture as the seasons change. These nomadic pastoralists can be found in environments as varied as the north of Scandinavia, where the Laplanders herd reindeer, to the dry grasslands of Africa where cattle are the main source of wealth for many groups. This nomadic life is not very secure. It depends on the luck of the weather in providing grazing for livestock. If the weather fails, then animals and people go hungry or even starve to death.

Apart from the pastoralists who herd animals for their own needs, there are some who do so on a commercial basis. There are still huge open ranges in Australia, for instance, where beef cattle are raised on scanty natural pastures before being sent to richer grazing lands for fattening up before being slaughtered. Far more cattle and sheep are nowadays raised on carefully managed grazing land, and many crops as well as grass are cultivated especially for domestic animals to eat. The animals convert the plants into a variety of foods such as meat and milk. Human beings then kill the animals and eat the foods they have made! The extreme form of this is 'factory farming' where the animals are almost like machines for converting food from one form into another. Many people are uncertain in their minds whether this sort of farming is right, and whether man should treat animals in this way.

Animals have also been hunted or kept for other resources such as skins, furs and bone. Peoples such as the Eskimo learned how to use every bit of the animals they killed. But just as many people are worried about factory farming, so are they about killing animals for their skins, furs or tusks – luxury items that no one really needs. On the other hand there seems little objection to using, say, the wool from sheep, since they suffer very little from the loss of their coat. Finally there are many examples of animals being used to carry people or goods. Sometimes the name 'beasts of burden' seem just right, but in many societies there are instances where men and animals such as dogs and horses show a respect and friendship for each other.

1. Look through this book and give an example of animals being used for **a)** a source of clothing, **b)** material for shelter, **c)** carrying people or goods and **d)** food. In each case name the animal, the place where the animals are found and the people using them.
2. Draw a simple 'food chain' to include cattle and man, with man at the top of the chain.
3. Do you think there is any difference between animals being killed for Eskimo clothing and for luxury furs? Do you think there is any difference between buffalo being killed by Red Indians for food and cattle or sheep being killed in this country for their meat? Is killing animals always wrong? What, if anything, makes it acceptable?

In this Argentinian meat factory, a chain saw is used for cutting up a carcass as if it were a log in a sawmill

In this huge warehouse in Australia wool is packed and exported to overseas countries

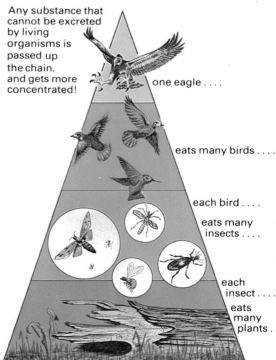

A 'food chain'. Food chains begin with plants, since these are the only organisms able to convert the sun's energy into chemical energy, which all animals and birds need

10 Animal resources

Twentieth century herders

The way of life of these Arab shepherds has changed little since Biblical times

Nomadic or semi-nomadic herding of cattle, sheep and goats is less important than it once was, but there are still some areas in the world where this sort of farming can be found. Usually they are where the climate is the sort to produce only seasonal grazing of grasses or scrub, or where the relief of the land is fairly rugged. The part of the Middle East known by Christians as the 'Holy Land' has always been a land of sheep and shepherds. Even today, with the growth of modern farming, industry and cities in Israel and its neighbouring countries, the traditional form of sheep farming survives. In fact this style of nomadic pastoral farming can still be found all the way from the Middle East through the mountainous areas of Iran and Afghanistan and into the Himalayan foothills of the USSR and Mongolia.

Keeping cattle is one of the main farming activities in Africa, although not everywhere, as the map shows. Modern methods of animal farming are replacing traditional forms, but these survive among some groups. The largest of these groups is the Fulani. These West African people graze their cattle on the natural vegetation of the savanna lands that lie between the desert to the north and the coastal rainforest to the south. The trouble is not only that rainfall decreases northwards, but that it is also seasonal and not completely reliable. So the Fulani move their herds north and south to follow the rainfall and fresh pasture it produces. They migrate north with the rains in April and May and south at the end of the rains in October. This can mean a yearly trek for families of up to 800 kilometres or more, with halts being made for short spells in temporary camp sites.

Above: This shepherd boy roams with his sheep over the hills of Afghanistan

Fulani on the move in Niger, where camels are more useful than horses

Their homes are made from guinea corn stalks, leaves, mud and other local materials. They carry these materials with them and any other utensils and possessions on cattle, donkeys or sometimes camels. In the drier season the daily pasturing may involve a walk of about fifteen kilometres from the campsite, but in the wetter season pasture will be closer at hand. Herds vary in size from about twenty to a hundred or so cows. Income is got from selling dairy produce – fresh and sour milk, butter and cheese – in local markets. The Fulani do not eat their own cattle – if meat is needed it is usually bought from a market.

The life of the Fulani is changing. Their clothing is now of cloth rather than the traditional leather, while plastic sandals, transistor radios and bicycles may be seen being used by Fulani who go to the larger markets. The nomadic pastoral life of the Fulani seems likely to end over the next few decades.

1. What are the differences in the pastoral farming shown in the photographs of Israel and Afghanistan – mention types of animal and environment, if possible. What might make intensive farming of crops difficult in both these areas?
2. What are the main differences in rainfall between the northern and southern savanna areas?
3. Why is cattle farming possible at A, but not at B or C on the map?
4. In what sense was the traditional Fulani method of farming practical and efficient? Why is it not very suitable for a developing country that wants to produce more food?

A Fulani family loading up before moving on

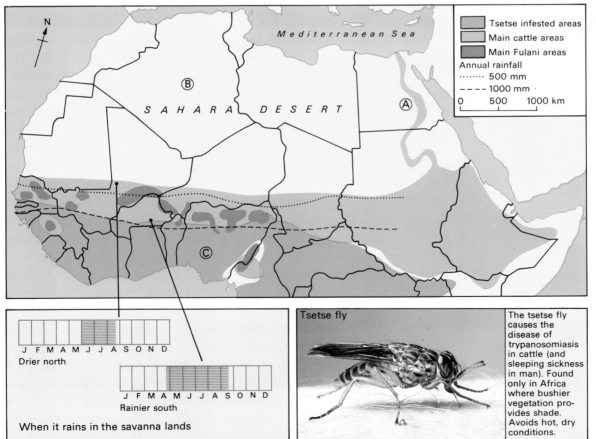

West African savanna lands

When it rains in the savanna lands

The tsetse fly causes the disease of trypanosomiasis in cattle (and sleeping sickness in man). Found only in Africa where bushier vegetation provides shade. Avoids hot, dry conditions.

10 Animal resources

Commercial animal farming

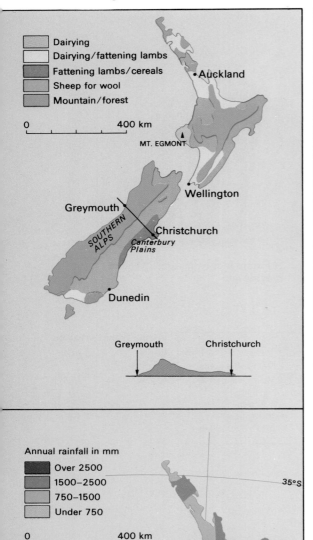

New Zealand. a) land use: types of farming, b) rainfall

Sheep being 'fattened up' on the Canterbury Plains, South Island, New Zealand

The wealth of New Zealand is based on its animal resources. Wool, lamb and dairy produce make up a large part of its exports. But the type of farming also varies within the country. The maps of rainfall and the cross-section help to explain why. The latitude of New Zealand is about the same as that of Spain – though south of the equator, of course! This suggests it will be quite warm at lower levels for most of the year. But there are very high mountains in much of the country – fold mountains such as the Southern Alps in South Island and volcanic peaks such as Mount Egmont in North Island. These are too high, rugged and cool for farming. The islands that make up New Zealand are a long way from Australia, the nearest large land mass, and so the north-westerly air masses bring a lot of rain to the country, particularly on the west coasts. This combination of climate and land produce excellent grazing lands or farmland on which animal foods can be grown. The grasslands are also carefully managed to improve their quality.

The farmers are able to produce far more animal products than the small population needs. Until methods of refrigeration had been developed, and special ships built, it was pointless New Zealand farmers doing so. But when this technical development did take place then New Zealand was able to sell large amounts of animal produce all over the world. The British were the first Europeans to settle New Zealand, and after fighting and overcoming the Maoris they made the country a part of the British Empire. This explains why trade between Britain and New Zealand has always been strong.

New Zealand is a little bigger than Britain, but its population of under three million is only about one twentieth the size. As the graph shows, the average income per person is quite high. The wealth of a country seems to have little to do with the size of population. On the other hand there does seem to be a link between average income and percentage of the population who are farmers – the higher the percentage, the lower the average income.

When the United Kingdom decided to join the European Common Market, she had to agree to certain changes in trade, especially of farm produce. New Zealand and Britain could not make special agreements. As a result New Zealand had to look for other markets and change its pattern of trade. Commercial farmers have to sell their goods somewhere or their personal income, and that of the country as a whole, will fall.

1. Copy the profile between Greymouth and Christchurch. Label with the following phrases: heavy rainfall – little farming; less rainfall – sheep farming for wool; low rainfall – cereal growing and fattening lambs.
2. Which type of sheep farming – fattening on the Plains or wool production and breeding in the foothills – is likely to have the largest farms? Say why.
3. What are the main differences between the type of cattle farming practised by the Fulani in the savanna lands of Africa and the New Zealanders on the lower grasslands of North Island?
4. From the graph, name an exception to the rule, 'the lower the percentage in farming, the higher the average income'.

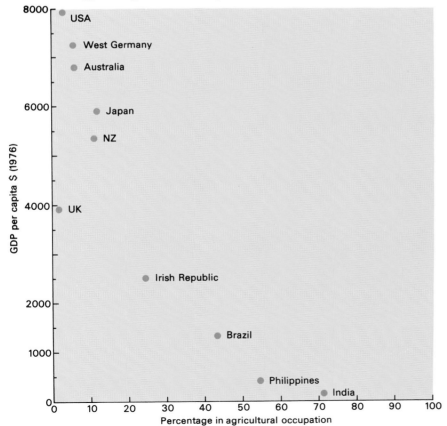

Trade between New Zealand and the United Kingdom
(as a percentage of total trade)

	NZ exports to UK	NZ imports from UK
1940	88	47
1950	66	60
1960	53	43
1970	36	29
1975	22	19
1978	18	18

New Zealand exports

	Percentage
Food/live animals	45.6
Inedible raw materials	29.2
Manufactured goods	10.9
Remainder	14.3

Below: A New Zealand dairy farmer near the Volcanic Mt Egmont, North Island

A large slab of New Zealand butter being removed from a stainless steel churn

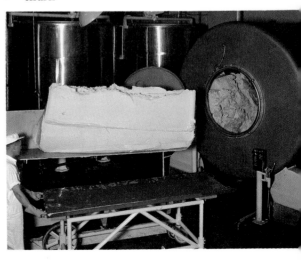

10 Animal resources

Animals at risk: who cares?

Right: The Prince's 'bag'. In one day's shooting in Nepal, the Prince of Wales (son of Queen Victoria) and his friends shot seven tigers

African poachers celebrate the slaughter of an elephant

This lion has been drugged by game wardens in a National Park in Namibia. Perhaps it was ill and needed treatment or had wandered out of the Park and had to be returned

An animal species becomes extinct when it fails to produce enough young in each generation to keep pace with the death-rate. We can tell from fossil evidence in rocks that many living species have become extinct over the millions of years since life began. It is a natural process and extinction is the fate of any animal that has specialised too far to change when its environment changes, or has to compete with a better-adapted and more powerful animal. Because of enormous technical developments during the past few centuries, man has destroyed or nearly destroyed some species by killing them at such a rate that they couldn't produce enough offspring, or by completely changing their natural environment at tremendous speed.

A number of examples have already been given of the way in which natural environments are being rapidly changed – Amazonia, for instance. There is every likelihood that many species of creature will be made extinct because of these and similar clearances of natural vegetation. The way in which animals have been hunted and killed for food have also been described. The North American Buffalo is a case of the near-extinction of a species through hunting. Often the numbers are so great the hunters may not realise the danger. But even when the danger is widely publicised, as with whales, the financial rewards for the hunters may be so great that they choose to ignore the threat to the species. Attitudes like this have led to hunters and poachers killing animals for furs, for ivory or merely for ornaments for tourists. A slight variation on this is when sportsmen and tourists hunt animals for trophies. Magnificent creatures such as lions and tigers (however ferocious and terrifying to man and other animals) have been hunted out of existence in some parts of the world. It is important to realise, though, that animals are sometimes killed out of fear. 'Killer tigers' are eliminated in this way. And animals are sometimes killed out of a wish to reduce numbers to help the species to survive. The killing of the Canadian seals and their pups is claimed to be for this purpose, and the use of their skins for luxury furs is only a by-product.

Many people are concerned about animals and wildlife conservation. One way to preserve species under threat of extinction – whatever the cause – is to remove them to zoos and parks and breed them in captivity. There is always the chance that enough offspring will be born to return them one day to their natural environment – provided it still exists, and that hunters and poachers don't begin to exterminate them again! Another method is to protect the animals in their natural environment by creating wildlife reserves and parks and using game wardens to look after them. But the parks are large, the wardens few and the determination of hunters and poachers very great. Early in 1980 wardens and poachers clashed in East Africa. The gangs of poachers were armed with modern weapons and several people were killed.

There is great pleasure and satisfaction in watching wildlife in natural or near-natural environments, and tourism can add to the income of countries. The animals are still resources – but in a very different form.

1. It is said that man kills animals for 'fear, food, finery (feathers and furs), fun and financial gain'. Look through this book and give one example of each of these. Can you think of any other reason why man might kill animals?
2. Man seems to have existed for several million years. Why is it only relatively recently that man has been such a threat to the survival of so many species?
3. Compare the scenes of the tiger hunt and the elephant poachers. Do you think that one is more justified than another? Give reasons for or against both.
4. What do you notice about the locations and environment of the endangered species shown on the map?
5. Do you think it sensible to try to preserve nearly extinct species in zoos, or would it be best to let them die out as many species have before?

Squirrel monkeys are much in demand as unusual pets by people in the West. This is how they are transported. Many, of course, do not survive.

Wildlife at risk: a few larger species

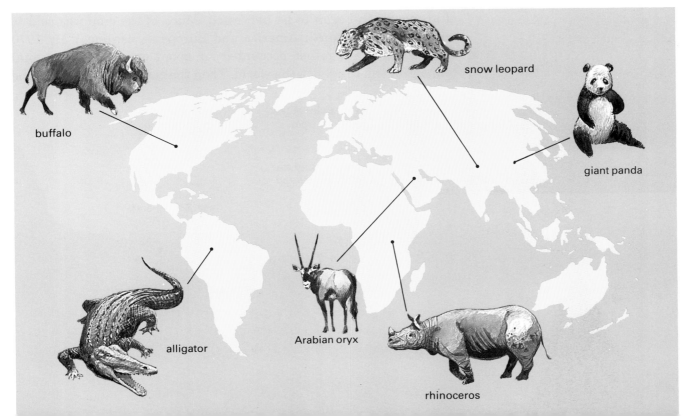

11 Resources from the sea

Fish: harvest of the oceans

This fisherman from Southern Italy is taking his fish home in an unusual way

Over seventy per cent of the earth's surface is covered with the water of the oceans and large seas and gulfs. But the surface of the ocean floor is as varied and rugged as that of the land. It ranges from the very deep ocean trenches to the shallow seas around the margins of the continents. These shallow sea floors are really parts of the continents, which is why they are given the name of continental shelf. Here and there the fold mountains and volcanic peaks create ocean ridges that sometimes rise above the level of the sea to form islands or island arcs. These ocean waters provide environments for hundreds of different species of fish.

At first it would appear that the oceans offer limitless resources of food for fish, but this is far from true. Only the top layers of the oceans that can be penetrated by sunlight can support plant life that provide fish with their food. The result is that there are 'grounds' where fish are – or were – abundant, and other places especially in the open sea where there are far fewer. Not all fish are suitable or wanted as food and it shouldn't be forgotten that many are caught by other creatures – larger fish, birds and whales, for example. Many methods of fishing are practised around the world, using lines or nets. Sometimes these are cast and hauled in by hand, but in modern commercial fishing very large vessels using machine-hauled nets are used.

One area where the waters provide plenty of food for fish is off the coast of Peru and northern Chile. This is because a cold water current full of nutrients flows past these shores. During the 1950s vast shoals of anchovy fish began to be harvested. Much of the crop was sold to meat farmers in North America and Europe as fish meal. By 1970 Peru and Chile together accounted for about one fifth of the total tonnage of fish caught in the world. Then the catches began to decline

Left: Work on board a trawler at the height of a storm can be far from pleasant

Below: Diagram comparing highest mountains, continental shelves and ocean depths

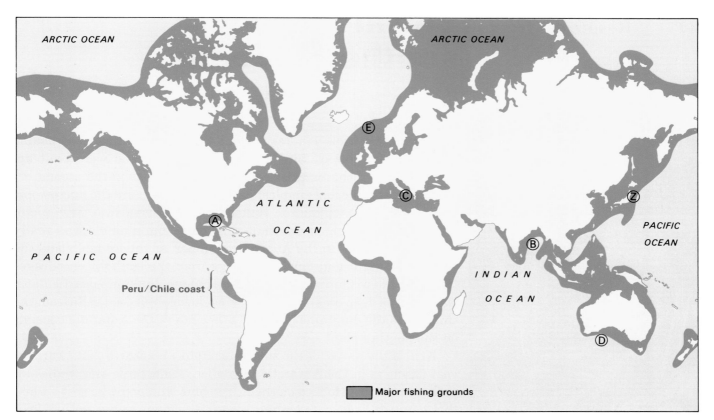

Oceans and major fishing grounds

dramatically. Not enough were left to ensure fresh stock were born. There is every sign that the anchovy have been over-fished. It is possible that some of the difficulty is due to the interference of the up-welling cold currents by a warmer counter-current from the north, but there is little doubt that the main reason is over-fishing. A resource that, properly managed, might have lasted for centuries has been virtually wiped out for short-term gain. To make things worse, anchovy is near the beginning of some food chains and its extinction would damage many other species.

Over-fishing is one way in which fishing resources can be destroyed. Another is if the seas become heavily polluted. This is most likely to happen where large rivers take farming or industrial pollution into the seas, or where large cities with rapidly growing numbers line the shores of shallow seas. The Mediterranean has provided its countries with many varieties of fish, but unless the controls on pollution agreed between the countries in 1980 are enforced, the Mediterranean sea will become 'dead' – fish will be unable to survive there.

A rich catch of anchovies. Peruvian fishermen are putting themselves out of a job by overfishing

1. Name the sea areas lettered A–E on the map. The country Z is one of the biggest fishing countries in the world. Which country is it?
2. Draw a graph to show the following changes in the Peruvian fish catch. The figures are in millions of tons.
   1957: 0.5   1960: 3.5   1965: 7.6   1970: 12.6   1974: 4.0
   Add a short note of explanation to your graph.
3. Show the following facts in a diagram. Ocean catches 1980: Pacific, 55 per cent; Atlantic, 38 per cent; Indian Ocean, 5 per cent; Mediterranean and Black Seas, 2 per cent.

11 Resources from the sea

Sea life in the southern oceans

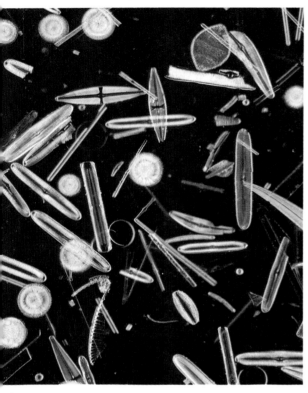

Phytoplankton, the microscopic plant food found in the world's oceans

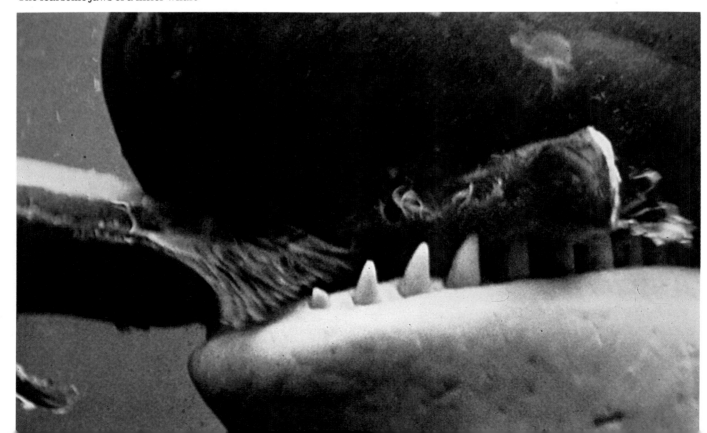

The fearsome jaws of a killer whale

Earlier in the book you read about the way in which food chains between plant, insect, bird and animal life on land were built up. Similar sorts of chains exist between living things in the oceans.

The most basic and simplest parts of this chain are the microscopic plants called phytoplankton that float near the surface. The microphotograph, though, shows that this 'basic' building-block is really quite complicated. In the Antarctic summer, as the ice melts and the sun shines night and day, the energy from the sunlight encourages the plant life to flourish. The Antarctic phytoplankton is particularly rich because deep ocean currents bring down plant-food salts from the tropics. As the diagram shows, all other life in the Antarctic depends on the floating plant life.

Krill is the name for the small shrimplike creatures that swarm in vast numbers in the Antarctic seas. Fish, seals, birds and whales all gather in the area to feed on them. The blue whale, the largest animal that has ever lived, eats nothing else and as many as two million krill have been found in the stomach of a single blue whale. Krill eat phytoplankton and other small floating animals. Their huge swarms, which may be up to 400 metres long, colour the sea pink. They live for two or three years and their lives are influenced by the movements of the Antarctic Ocean currents. They spawn in late summer and lay their eggs in the cold, north-flowing water. The eggs sink to the

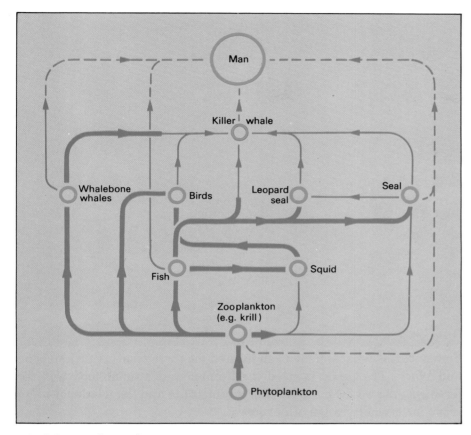

A food chain in the southern ocean

15 Nations agree to protect fish of Antarctic

Canberra, May 20 Fifteen countries today concluded an agreement on conserving fish stocks in the Antarctic, now the target of increasing numbers of trawlers.

The nations, which included the United States and Soviet Union, agreed on a convention to prevent over-fishing by regulating commercial exploitations of marine life. An international commission of experts will be established to study the food-chain of Antarctic fish and bird life and recommend measures to protect species.

The two-week conference of the 15 nations was concerned with the increasing exploitation of the shrimp-like krill, a high-protein crustacean, which has a pivotal role in the food-chain.

The countries that reached agreement today were East and West Germany and the 13 Antarctic treaty members – Argentina, Australia, Belgium, Britain, Chile, France, Japan, New Zealand, Norway, Poland, South Africa, the Soviet Union and the United States.

An extract from *The Times* of 21 May 1980

A trawler catch of dead krill and the fish that live on them from the rich waters of Antarctica

warmer south-flowing water, where they hatch. When the larvae mature into krill they swim to the surface.

Krill are eaten by many kinds of fish, penguin and other birds, squid, seals, and baleen whales. The larger species feed on each other – birds and seals eat fish, toothed whales eat penguins and seals as well as fish. The killer whales are feared by all the other creatures. They will even attack other whales. They are curious and intelligent, and move about in packs. They are found in most oceans of the world. A pack turns up, terrorizes the neighbourhood, and then departs as suddenly as it arrived.

Man has now appeared on the scene, hunting all these creatures. He has almost exterminated some species of whale and is in danger of upsetting the natural balance of life in the oceans as more and more krill are fished by more and more countries. The extract shows how some attempt is being made to control this possible devastation of life in the southern oceans.

1. Put into words the food chain that links phytoplankton and whale meat sold as food to humans. Follow the longest route shown on the diagram.
2. List the countries named in the extract into two groups – those with coasts in or near the southern oceans and those in the northern hemisphere. Why are countries in the second group fishing in the area? Have they any less right than those in the first group?
3. Why is it difficult for countries to agree on a maximum catch to avoid over fishing? Why is it difficult to make sure that everyone keeps to the agreed limit?

11 Resources from the sea

Hunting the whale

Right: An old engraving of an early whale-hunt off the coast of Newfoundland

An extract from the *Sunday Times* of 4 May 1980

Whalers admit: We break law

by Brian Jackman and Tana de Zulueta

Spain is grossly overfishing its official quota of whales. Following last week's sinking by saboteurs in northern Spain of two 500-ton whaling vessels, the owner of the ships admitted that his company had killed far more whales than allowed under the quota.

The vessels, the *IBSA Uno* and *IBSA Dos*, were alongside the jetty with their sister ship *IBSA Tres* at Marin, a small port in the Pontevedra estuary.

Shortly before two o'clock last Sunday afternoon a series of explosions ripped through the two ships, instantly sinking them. Normally, each ship had a 13-man crew, but no-one was aboard at the time. The *IBSA Tres* is now under armed guard.

No one has yet claimed responsibility for the attack – which has effectively destroyed half the Spanish whaling fleet – but it was most probably the work of an extremist conservation group. British anti-whaling groups were at pains to deny any involvement.

In the past two years Spain's whale catches have quadrupled because of Japan's growing demand for whalemeat.

Greenpeace, the conservationist organisation, says the Spaniards have admitted killing blue and humpback whales, which are nearly extinct, whenever they find them. The owner of the sabotaged whalers, Juan Masso, denies this, but he told the *Sunday Times* that last year his ships killed more than 400 whales. Spain's quota for this year was 143 whales.

For centuries whales have been hunted for their meat, oil, whalebone and ivory. The meat is used for human and animal food and for fertilizer, the oil for margarine and lubricants and the whalebone and ivory for a whole range of purposes.

The whaling industry proper began in the seventeenth century with British, Dutch, Norwegian and American whalers hunting the whales that came into the coastal areas of the Arctic to breed. Open sea whaling began in the following century, and Japan became one of the leading nations involved. Ships sailed regularly to the coasts of Australia and New Zealand, and soon the southern Atlantic, Pacific and Indian Oceans within the Antarctic became very important. The period of modern whaling began in 1864, with the invention of the harpoon gun. This, together with the use of steam power, made the hunting of the faster and more manoeuvrable whales such as the blue fin, fin and sei, a practical possibility. The faster steam-powered whalers could penetrate the pack ice where these species gathered in their millions. Steam power enabled the catchers to inflate the carcasses of the dead creatures to keep them afloat while towing them to the factory. Modern vessels are powerful enough to tow several at the same time. Until the late 1920s the Antarctic whaling industry was based on land factories, mainly on South Georgia and the South Shetlands. Then factory ships appeared in Antarctic waters, and the meat and oil could be obtained at sea and any waste dumped overboard. The whaling fleets became completely mobile. In addition to the factory ship, they consisted of refrigerator and transport ships for storing the meat, oil and other products and perhaps a dozen catcher ships. Helicopters or light spotter planes are nowadays used, and the fleets can follow the whales wherever they go.

The modern harpoon has an explosive shell in its head and is fitted with hinged barbs that prevent its withdrawal after penetrating the whale.

The Antarctic whaling fleets have almost destroyed the resource that they depend on! In spite of warnings from scientists over the past sixty or seventy years, they have plundered the stock of whales wherever they managed to catch them. Some countries have now ceased whaling altogether, others forbid the import of parts of the animal into their country, and most agree to quotas set by the International Whaling Commission. Unfortunately the body has little real power to stop countries breaking the quota, as the extract shows. Blue, fin and humpback whales have been hunted almost to the point of extinction. It has been calculated that even with complete protection it could take some fifty years for the humpback stock to recover – but if the past is anything to go by, the greed of some is not likely to allow any species to remain unmolested in order to survive!

The Greenpeace Foundation is concerned with stopping illegal whaling. Its boat acts as a 'watchdog' on the high seas

1 From a globe or atlas, give the latitude and longitude of South Georgia, the South Shetlands and the Bering Strait. (See also Bowhead whales, pages 40, 41.)
2 Choose one of the two illustrations and describe what is going on. Imagine you are one of the people involved. Say which you are, and what you are thinking about the activity going on.
3 What are the main differences between the whaling practised by the Eskimos of Alaska and the commercial whaling fleets?
4 Imagine you are either **a**) and shipowner or **b**) one of the people who blew up the ships at Marin. You have been given one minute of radio time to justify your actions. Prepare your own script. What would you say?

Whale catch quotas imposed by the International Whaling Commission

	1978/79	1979/80
Sperm	9360	2203
Minke	9173	12006
Sei	84	100
Fin	455	624
Bryde's	454	143
Grey	178	179
Bowhead	18	18
Humpback	0	0
Total	19722	15883

Japanese whalers cutting up their catch at a South Atlantic whaling station

11 Resources from the sea

Changes in British fishing

A group of old-fashioned fishing boats in Lerwick, Shetland

A fish auction in Hull

The seas off North-West Europe provide many of the sorts of fish that are in great demand for food in this country. The combination of shallow seas over the continental shelf and the continual mixing of ocean currents means that these waters are rich in fish foods. Apart from the lobsters and crabs, cockles and mussels, shrimps and prawns found in estuaries and very near the coasts there are two main sorts of fish. There are those such as cod, haddock and the flat-fish like sole and plaice that swim near the sea bed. Then there are those such as herring and mackerel that live near the surface of the water. Different methods of fishing are used to catch these different varieties.

The smallest boats tend to fish the so-called in-shore and near-water grounds, while the larger ships go further to the middle-distance and deep-water fishing areas. These deep- or distant-water grounds are off Iceland, northern Norway, Greenland and Canada, and a trip from a British port takes three or four weeks. That is why the development of modern navigation and large freezer-trawlers became important several decades ago. Nowadays some ships are really floating factories that do no catching. They receive their fish from smaller vessels. A lot of the fish is taken straight to fishing ports, such as Hull and Grimsby, where special equipment and factories handle the catch.

There has been a dramatic decline in the British fishing industry in recent years. In the mid-1970s for example there were over 500 trawlers in Britain's deep sea fishing fleet. By 1980 there were only about 130. One reason for this is that most countries have now established a 200-mile fishing zone around their coasts and allow only limited fishing by foreign vessels within it. So the once important distant water grounds can no longer be fished by British ships just as they please.

Another reason why ships and owners are finding life difficult is that there are now restrictions on the catching of some varieties of fish in waters nearer Britain and even a complete ban on others. Because of over-fishing, for example, there was a danger that herring would completely disappear. So for conservation reasons there is a strict control on herring fishing. Mackerel fishing is also restricted to certain months. A combination of modern, efficient methods of fishing and the wish to make as much money as possible has led to the near destruction of some fish resources. The fishing industries of some neighbouring European countries are, unlike Britain, given a lot of financial support by their governments. Because they belong with Britain in the European Community they can fish the North Sea and sell their catch to British buyers far more cheaply than British ships can. So the British owners and fishermen lose money and finally go out of business. Ports like Hull, the biggest fishing port in Europe in the mid-1970s, were facing a complete collapse of their fishing industry in the early 1980s.

1. Draw simple diagrams to show the following information.
   a) Trawlers in Hull: 1976: fresh fish, 55; freezer trawlers, 40.
   1980: fresh fish, 4; freezer trawlers, 25.
   b) Fish landings at Hull: 1955: 250 000 tonnes.
   1976: 110 000 tonnes.
   1979: 50 000 tonnes.
   Write a sentence saying what the diagrams illustrate.
2. Make a larger copy of the central part of the profile on page 88 showing the shore and continental shelf covered with a shallow sea. Add to it labels or drawings to show where mussels, crabs, shrimps, sole, cod, herring and mackerel are likely to be found.
3. Choose one of the illustrations of commercial fishing from this page or page 88. Describe the scene and write an account of how you would feel about working on one of the ships.
4. Do you think that foreign fish landed at British ports should be more heavily taxed (so increasing the price in the shops) in order to help British fishermen to sell their catch?
5. Do you think our taxes should be used to support British fishermen or should the British fleet be allowed to disappear so long as we can get cheaper fish from other countries?

An unemployed trawlerman leaves his boat for the last time at Fleetwood in Lancashire, where the number of trawlers has fallen from 150 to under 20

Continental shelf and distant water ports

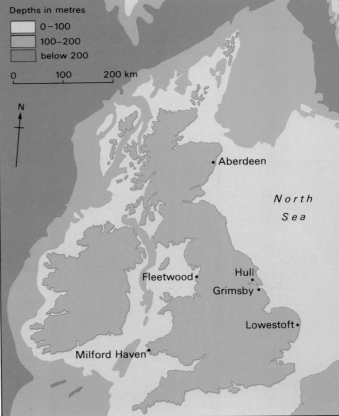

Left: A protest in London by trawlermen against the decline in their fortunes

12 From the earth's crust

North Sea oil

Oil men returning to a rig after one of their regular periods of shore leave

Right: Oil from the North Sea

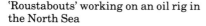

'Roustabouts' working on an oil rig in the North Sea

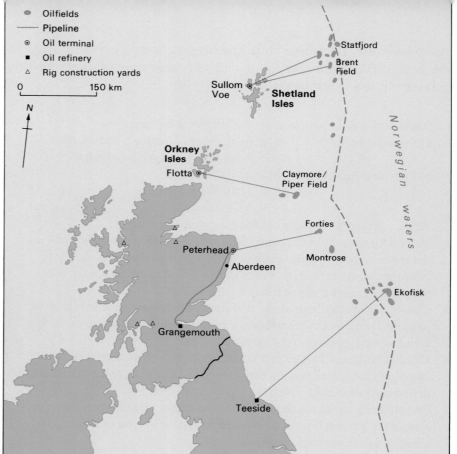

Some of the rocks that make up the earth's crust are extremely valuable. Amongst these are those that provide the energy – coal, oil, natural gas – needed in the modern world. These resources are the end product of the breakdown of dead vegetation by bacteria and the sheer pressure of being buried under layer upon layer of sediments.

Oil is not found in massive underground pools or reservoirs, but trapped along with natural gas and water in the tiny spaces in certain porous rocks. For this to happen the porous rock must be sandwiched between non-porous rocks and either folded or faulted in such a way that the oil and gas can collect in 'fields'. Because these fields of oil are often deep underground, it is not surprising that it is only recently that many have been found. Oil is so vital to modern life that vast sums of money are being spent on exploration and test-drilling to find new reserves. Often the discoveries are made in places with very harsh environments – or places so attractive that it seems wrong to damage them by oil development. Even when discovered, the reserves may be so remote or difficult to get at that it is not worth trying to recover the oil.

About 150 kilometres north-east of the Shetland Islands, beneath the stormy, grey North Sea, lies the Brent oil field. It is one of the larger of a number of fields discovered in the early 1970s in the seas controlled by Britain. The trouble is the oil-bearing rocks are beneath some of the roughest waters in the world, and the huge drilling rigs and production platforms are exposed to strong often bitterly cold winds. Brent A, for example, stands four times as high as Nelson's Column. Even when the oil is recovered, it has to be transported to the mainland refineries. It is only because oil is essential for survival at the moment that vast sums of money are spent on this extremely difficult job.

To recover as much oil as possible, many separate holes are drilled and water pumped down some of them to maintain pressure in the field as the oil is extracted. In the Brent field, natural gas is produced as well as oil.

In some North Sea fields the oil is pumped into a storage buoy and transferred to tankers which take the oil to the mainland refineries. These storage buoys can hold over 40000 tonnes of oil. In the Brent field however the storage buoy and tanker system has been replaced by an underwater pipeline linking the platforms to a giant oil storage terminal at Sullom Voe on the Shetland Islands.

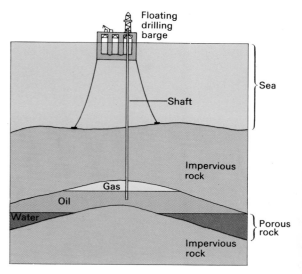

A simple rock-structure containing oil and natural gas

1. Give the name of the main production field and the terminal of each of the five pipelines shown on the map. What is surprising about the position of the most southerly of these four fields?
2. Imagine you are writing a letter to a friend who has never seen or even heard of one of these oil platforms. Explain what they are like and what life is like as a worker on one of them.
3. What are likely to be some of the effects on a) the people and the environment in certain parts of Scotland, b) the wealth of the whole United Kingdom, as a result of the discovery and recovery of oil from under the North Sea?
4. Oil and natural gas are amongst the so-called 'non-renewable' resources. Try to explain what 'non-renewable' means.

An artist's impression of two different kinds of oil rig, showing how they can be fixed to the ocean floor

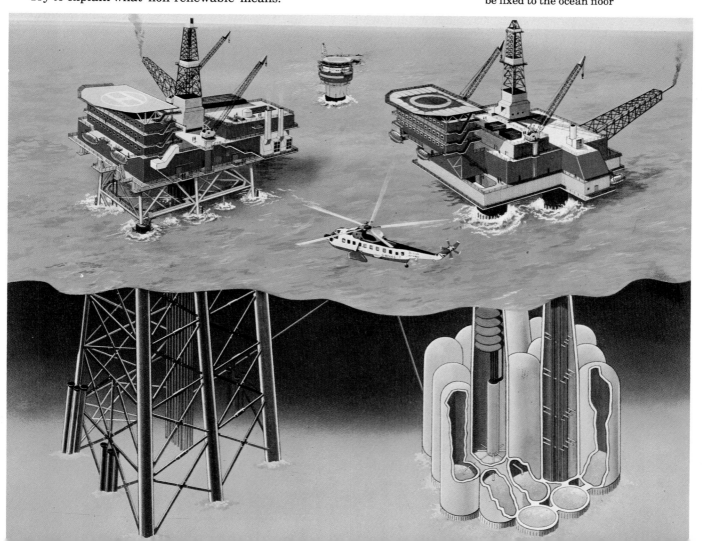

12 From the earth's crust

North Sea oil: gains and losses

How the North Sea oil industry has changed Sullom Voe in the Shetlands. Late 1978 (*left*) and early 1981 (*right*)

Below: a) UK oil refineries. b) Canvey Island

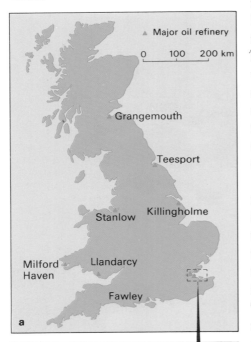

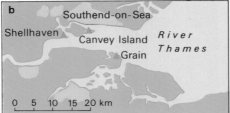

The discovery and the use of North Sea oil has had a mixture of good and bad effects. There is no doubt that without the wealth provided by the oil the people of Britain would be far worse off. It has meant that large sums of money have not had to be spent on buying oil from foreign producers, and some has been earned by selling British oil. It may seem strange to read that Britain needs to import certain varieties of oil while exporting others until it is realised that there are different sorts and qualities of crude oil. The good fortune of having oil and natural gas resources of its own, and the skill and ability to recover it, means that Britain has a few years to try and find alternative sources of energy before oil runs out – or becomes too expensive to use.

The country as a whole benefits from North Sea oil. So also do many of the communities in Scotland – although many claim that Scotland is not getting its fair share of benefits. Sullom Voe oil terminal in the Shetland Islands cost hundreds of millions of pounds to build and is the largest in Europe. Before the oil companies went to the Shetland Islands, the islanders lived quietly from fishing, crofting, sheep tending and a cottage industry based on knitted woollens. The islanders are estimated to be likely to get between 50 and 100 million pounds before the year 2000 from its agreements with the oil companies. It has not been all that easy for the Shetland islanders to decide how to use all this wealth – there are only about 20 000 people on the seventeen inhabited islands. Some 1 500 local men and women joined the 5 000 workers that built the terminal, and their lives were greatly changed during this time. Young girls working as waitresses and chambermaids in the hostels, for example, were able to earn more money than their parents ever had.

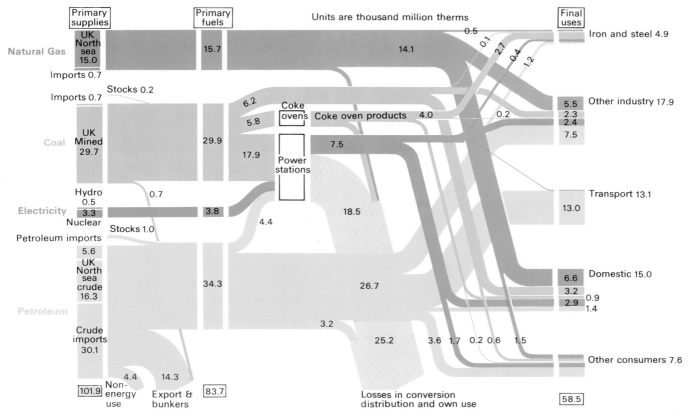

Energy flows in the United Kingdom, 1980

Huge supertankers of 300 000 tonnes use the terminal daily, taking the crude oil to refineries on the mainland. Although the terminal is carefully designed, and only a few hundred permanent staff and their families will come to live on the islands, there is no doubt that it will have a big effect on the people and the environment. It is interesting to know that part of the agreement between the islanders and the oil companies was that when the oil has been recovered soon after the year 2000, and the terminal is no longer needed, the tanks and buildings will be cleared away and the area restored to the rural scene it once was. The money earned can then be used to support future life on the islands.

Not all people and environments are so fortunate, though. Oil refineries where the crude oil is turned into petrol and many other products are large and very complicated works. They are also very dangerous if great care is not taken over safety. The people of Canvey Island, for example, are concerned about the proposal to enlarge the local refinery. The country needs the oil products, and local people need work – but the cost may be very high!

1 From the diagram work out very roughly what amount of energy at that time was provided by petroleum (oil) and natural gas. Why is it important to stress 'at that time'?
2 Draw a diagram showing the uses of petroleum – at that time. Arrange your diagram so that the uses are in some sort of rank order, with the largest first.
3 Imagine you were a villager living in the Shetlands before the building of Sullom Voe. What do you think you would like and dislike about the changes that had happened? How would your age be likely to affect your feelings about the changes?

The Times 7 June 1980

Expert's warning on Canvey Island

Flames would rise 3 000 ft above Canvey Island, Essex if the 380 000 tons of oil stored there ignited, American Professor Janus Fay said yesterday at a public inquiry on the island. Residents are trying to prevent the building of an oil refinery by United Refineries.

The very great risk of group accidents borne by the 34 000 population of the island is nearly 10 000 times greater than that for the rest of the United Kingdom. The proposed installation would add 20 per cent to the existing risk, he said.

He added that a spill of 5 000 tons of petrol from a shipping accident at one of the jetties on Canvey Island would cover 40 acres of water surface, burning 14 tons a second for 10 minutes and engulfing neighbouring jetties.

Neighbourhood shelters, which would be proof against a blast wave, fire, flame radiation and toxic gas, should be constructed so that all islanders were within a few minutes walk. But he doubted whether warnings could be given in time.

Moving the world's oil

12 From the earth's crust

The Alaskan oil pipeline, which runs from Alaska's north coast to south coast over 1000 km, stretches out into the distance. The raised pipeline enables caribou herds to migrate across it

World production, consumption and movement of oil in the late 1970s (remember that these figures change every year!)

Many of the world's largest known reserves of oil are in countries that do not need it all for themselves. On the other hand most of the industrialised or 'developed' countries usually have far less than they need for their industries, cars, lorries and aeroplanes. As a result there is a massive movement of oil around the seas and oceans of the world. It is mostly crude oil that is moved as it is cheaper and probably safer for the refineries to be in the importing countries. The cheapest and safest method of moving this crude oil is by large supertanker.

During the decades between 1950 and 1970 the apparently unlimited cheap oil from the huge Middle East fields helped to create a huge growth in economic activity in Western nations. Soon the United States was importing about a quarter of its needs from the area, while Western Europe imported sixty per cent and Japan one hundred per cent of their needs. Then from 1973 the ministers of the countries exporting oil (OPEC) increased the price of their oil about four times. It was not simply greed that made the producers put up their prices so sharply. The enormous increase in demand for oil meant that the oil reserves were being very rapidly exhausted. At that time it was predicted that supplies would completely run out within fifty years. Oil is often the main source of wealth for the producing countries, so it was hardly surprising that Saudi Arabia, Kuwait, Libya, Iran and Venezuela introduced programmes to conserve their supplies for the future.

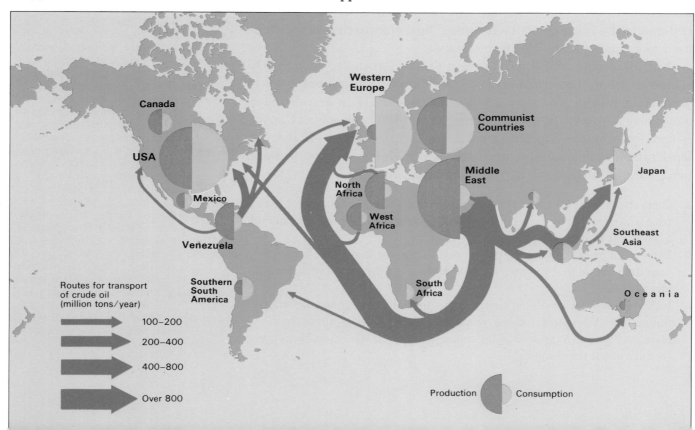

An oil tanker, the 250 000 ton *British Patience*

The oil tanker *Amoco Cadiz* lies broken, off the coast of Brittany. This single accident in 1978 covered hundreds of kilometres of beaches with thick oil

The main importers of oil – Western Europe, Japan and the United States – now realise they are threatened by this dependence on oil provided by other people. In times of trouble the producers could cut off this vital resource. Even if this did not happen it is now realised that there will come a time when there will not be enough oil for everyone. This will happen in spite of the discovery of enormous new fields such as those in Mexico. So the industrial nations are hurriedly seeking and developing alternative sources of energy such as coal, nuclear, solar and tidal energy.

In the meantime vast quantities of oil are being shipped around the world in giant supertankers. Every now and again an accident occurs to remind people of the disasters that can occur when handling the resource. A crash or spillage in coastal waters can cause chaos on beaches and destroy bird and fish life, while an explosion on a supertanker at a terminal would create devastating effects on the environment and people living and working nearby.

1. In which two areas shown on the map was the need very much greater than production, and in which four areas was production much greater than need?
2. a) If you were a country with a great need for oil, with which countries would you try and stay on good terms? b) If you had the means to do so, and wished to create chaos in Western Europe or Japan by stopping its flow of oil, which sea area or sea areas would you try to control?
3. Look at the photograph of the Alaskan oil pipeline and the information on page 44, and then at the map on the left. What *seems* to be missing from the map? Can you think of a reason why the map may not be wrong after all?
4. What are the advantages of moving crude oil in large supertankers rather than in smaller vessels? What are some of the problems and dangers?

This bird is a victim of pollution from an oil slick

12 From the earth's crust

How long can our minerals last?

This 'open-cast' copper mine in Utah is the result of a determined search for a scarce mineral

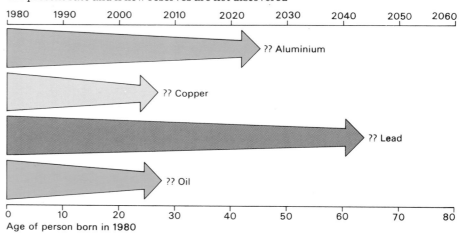

The date at which some minerals are expected to run out if they continue to be used at the present rate and if new reserves are not discovered

Age of person born in 1980

The crust of the earth contains vast amounts of valuable mineral resources that have been forming since the world began. Man slowly learned how to use some of them, and over the past few centuries has mined and quarried increasing amounts. There seemed no limit to the iron, lead, nickel, bauxite, coal and oil (among others) that was needed for modern industrial life. It seemed that given time and money, new sources would continue to be found. But in the past few decades people have begun to think differently. As the amounts we use have speeded up it is beginning to look as though very soon there will be no known deposits of some important minerals left.

There is no doubt that new discoveries will be made. In recent years, for example, oil has been discovered on a vast scale in Mexico, while it is now thought that there are huge reserves under the rainforests and Antarctica. It is now known that parts of the sea floor are covered in mineral-rich 'nodules'. These have been formed from the molten rock or magma at some depth in the earth's crust. In places, particularly the mid-ocean ridges, these deposits are actually on or close to the sea bed. All these examples show, though, that it is one thing to know that reserves exist. It is quite another matter to have the technical skills to recover them. Even when it is technically possible, the cost of doing so may be so great that at the time it just isn't worth it.

The minerals needed for modern industrial life are not evenly distributed around the world. Some countries such as Britain have to get many of their needs from others. All the chromium and nickel, for example, has to be imported into Britain. Unless deposits are discovered in Britain this will always be so. On the other hand some countries are very rich in mineral deposits. One reason for the wealth and strength of South Africa is that it is the biggest producer of quite a few minerals, as well as having the largest reserves of many. While

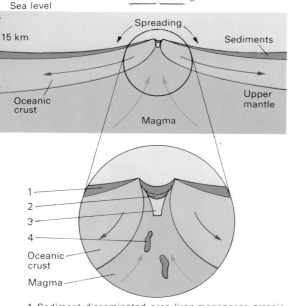

Below: The location of metal minerals on the mid-ocean ridge

1 Sediment disseminated ores (iron, manganese, arsenic, boron, zinc, lead, mercury, copper and iron)
2 Lens ⎫
3 Keel ⎬ massive ores (chromium)
4 Pod   massive ores (chromium, nickel, sulphide, platinium)

102

this is so many countries will remain dependent on South Africa for vital needs. Not all mineral exporting countries are so wealthy, though, especially if they depend on one mineral only.

It takes millions of years for new minerals to be created, and to all intents and purposes, when minerals are dug up and used they are gone forever. They are said to be non-renewable resources. At long last people are realising it is foolish and short-sighted to waste non-renewable resources. There needs to be a much more careful use of the resources being recovered at the present time. At the same time we should be exploring for new reserves and developing methods of using the known ones more fully.

1. Why is it difficult to be really sure when any particular mineral will be more or less used up? If the estimates are right, how old will you be when a) oil, b) aluminium ore (bauxite) is used up?
2. Look back through this book and at the photographs on this page. What are some reasons why exploration for and recovery of minerals is often very difficult? What is the name of the specialists who are paid by large companies to explore for minerals?
3. Mining is nowadays very big business, and some of the companies are 'multinationals'. What do you think a multinational company means? Why is modern mining such big business?
4. What are the dangers of one country being dependent on another country for a vital mineral? What can be done to avoid this dependence?

In the centre is copper as found in its pure state in nature. Surrounding it are products using this important metal

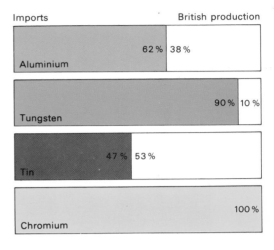

Four minerals in two countries.
a) Britain (*above*), b) South Africa (*below*)

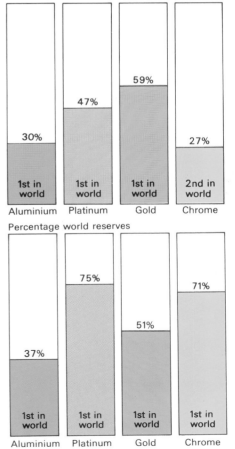

13 Human skill — Manufacturing things from resources

The *Mary Rose* (*above*), and an RAF Tornado (*right*), both technological marvels of their times

It is often said that the most important resource a country has is the skill of its people. From the earliest times man has used resources provided by the environment. Men and women have used their skills to turn crops, animals, timber and metal ores into more useful or valuable things, and used fuel and energy provided by coal, wind or running water to help them do so. Sometimes the skill of the craftsman produces something not only useful but beautiful as well. The value of many items of pottery, jewellery or furniture or buildings such as churches depends on skilled craftsmanship as much as on the use to which they are put. Simpler, cheaper and more easily made things would do just as well.

Very often, though, things are manufactured because they are needed for particular uses. This might involve just one resource and one person using human energy or muscle power – making an axe-handle or small boat from local timber, for example. Most manufacturing is a joint effort, where many resources are gathered together in one place and many people share their skills to shape or transform them into something different. The *Mary Rose* was built in 1509-10. She was one of the first purpose-built sailing warships, a technological wonder of her day and the pride of Henry VIII's navy. Many resources were used in building such ships – various timbers, sisal and hemp for rigging, cloth for sails, bronze and iron for the guns and pitch for caulking the timber to make the ship waterproof. Very great skill was used by hundreds of craftsmen in naval dockyards on the estuaries and coasts of Britain. The *Tornado* is another technological wonder, this time of the late 1970s. This advanced military aircraft is the result of remarkable skills being used to shape and assemble hundreds of thousands of parts into a spectacular machine. The resources, skills and methods of construction are vastly different from the *Mary Rose*, but the basic process of manufacturing is the same.

Manufacturing is likely to take place in the most suitable location. Usually, but not always, this is where costs of assembling resources, power, and workers and of transporting manufactured goods is lowest. The huge steelworks near Duisburg illustrates this very well. The River Ruhr is a tributary of the Rhine in Germany. The area through which it flows is underlain with rich deposits of coal suitable for making into coal for steel works. There is not much iron ore, so vast amounts are brought to Rotterdam at the mouth of the Rhine from Sweden, Liberia, Australia and elsewhere, and then taken by barge to the steelworks on the Rhine and Ruhr. Local coal is used with imported ore and scrap steel. The Ruhr is a fantastic mixture of coal mines, coke ovens, blast furnaces, steel mills and engineering workshops in a string of towns and cities stretching some sixty kilometres eastwards from the Rhine. The main reasons for the location of this great industrial area is the existence of the coalfield, an excellent river, canal transport – and the skill of the people.

1 Name one resource, one source of power and one skill used in making the *Tornado* that was not used in making the *Mary Rose*.
2 Why were most naval dockyards located on river estuaries or at some point on the coast?
3 a) Why is a waterside location very suitable for an iron and steelworks?
 b) Look at the map on the right. Through which huge port and which country does the iron ore have to pass to get to the Ruhr in Germany?
4 Choose any fairly simple manufactured item that you know. Name the resources or raw materials used in its manufacture and try to describe the way it is made. If you can, say where the item is manufactured.

One of the most important resources in the microelectronics industry is the skilled workforce

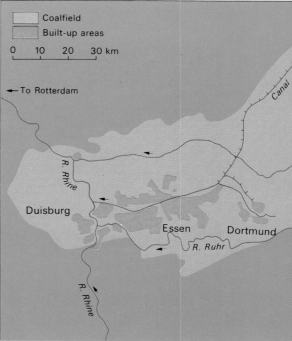

The Ruhr
Left: An industrial scene in the Ruhr Valley

13 Human skill — Making motor cars

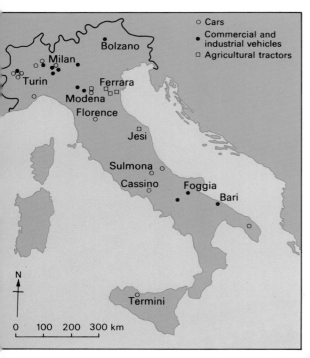

The location of Fiat car factories in Italy

The Fiat assembly line

Of the hundreds of thousands of manufactured products, few can have had more impact than the cars, lorries and coaches made by the automobile industry. Without motor vehicles life as we know it would come to an abrupt and devastating halt.

A modern car factory is usually organised to achieve the greatest production for the lowest cost, and this almost certainly involves mass production methods. In such a factory each employee specialises in a particular job such as designing, welding, fitting, operating a lath, handling accounts, managing production and testing the cars. A common method of production is by use of an assembly line. The different parts of the car are made in specialised workshops or perhaps in factories in different parts of the country. The parts or components are assembled as they move along a carefully controlled series of conveyor belts. An employee will be responsible for a particular job. When it is done the car moves on to the next stage and another car arrives for the task to be repeated. It can be a very efficient method of manufacturing cars, but can also be a very dreary and boring routine for the employee. In some car factories there has been a partial return to teams of workers building whole cars to give greater job satisfaction. Another development has been to automate production, using computer-controlled equipment to do the routine tasks, as shown below. The result of this, however, is that it will mean fewer jobs and probably greater unemployment.

Some factories, as we have seen, specialise in making components such as gearboxes, tyres and electrical parts for the assembly car works. These may all belong to the same company, or be independent and sell to a number of car manufacturers. This can lead to greater efficiency, but a breakdown or strike in a component factory can hold up production in the assembly works. On the other hand a closure of an assembly works can put many people in components factories out of work.

Motor vehicles are made in most leading industrial countries, and there is intense competition between the different companies. All try to increase their sales by exports. The result has been that many older and less competitive firms have had to close down because they could not sell enough of their vehicles. Another result has been for some firms in different countries to collaborate in various ways – British Leyland with the Japanese Honda Company, for example. It looks as though as time goes on there will be fewer and fewer different car companies and a few giant multi-national companies selling throughout the world.

1. Name four raw materials that are made into components that are then assembled into motor cars.
2. Name one advantage and one disadvantage of buying **a)** a mass-produced car **b)** a workshop-made car.
3. Describe what you think you would like or dislike about working on an assembly line.
4. Name one German, one Japanese, one American, one Swedish, one French and one Italian company that sells cars in Britain.

This picture shows a specialised Fiat-owned factory producing parts that will later be assembled into a car in another factory. You can easily tell that it has been built in the countryside

Even in the most automated car factory, there are still jobs for people to do. Here a man is knocking out dents in the car bonnet with an old-fashioned hammer

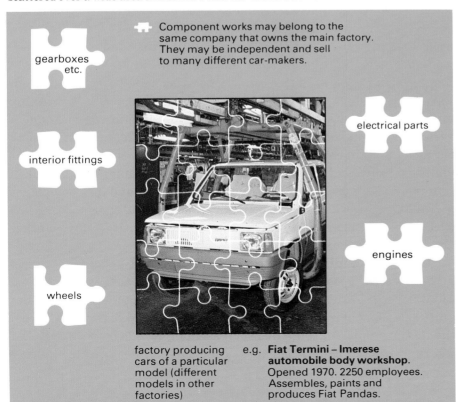

Several specialised component factories 'feed' the car assembly plant. They may be scattered over a wide area and sell to other car-makers as well

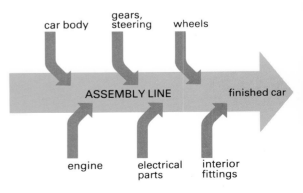

The order in which different parts are put together on a car assembly line

13 Human skill — Intermediate technology

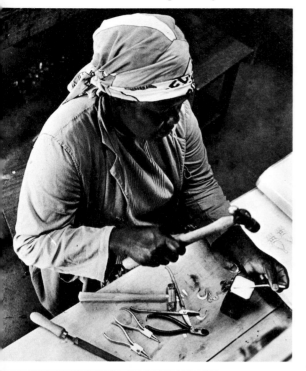

Making jewellery for the tourist trade, in a small workshop in Kenya

The dye pits at Kano, Northern Nigeria. The colourful pieces of cloth produced here are often sold overseas nowadays

This wind-driven generator has been built in a small village in Southern India at the cost of a few pounds

Not all manufacturing can be, or need be, as complicated or large-scale as the modern motor car industry. In the less industrialised countries a lot of people who could be classified as 'industrial workers' do not work in large factories, but own small businesses in their own backyard. In many ways they are not unlike the thousands of small workshops and lock-up garages that can still be found in Britain – though their numbers in this country have declined in recent years. In these small sheds and workshops wooden furniture, clay cooking utensils, straw hats, simple farm tools, paraffin lamps, sandals and a variety of other goods are made from resources such as local timber or old furniture, tin cans and car tyres. All these may seem insignificant and for purely local needs when compared with the motor car. But in countries where people have little money to spend and unemployment is high these workshops produce cheap and useful goods as well as jobs. Sometimes the crafts and goods produced reach a wider market and may even be exported for sale.

There are millions of people in the world today who need manufactured goods of the sort produced by the industrialised countries who either cannot afford them or find that the product is not quite right for their needs. They cannot even afford to buy goods from the newer industrial countries such as Hong Kong, Taiwan or Korea where things are usually cheaper. They may not be willing to receive 'aid' from countries such as America or China. In any case the complicated machinery is often not really suitable for their environments and when it breaks down they are unable to repair the fault easily or lack the necessary spare parts.

While these countries very often lack the money to start and run massive factories, a few of them are now showing what is possible. Using examples and help from elsewhere they manufacture products designed for their own special conditions from local resources and using their own skilled workers. In Tanzania in East Africa, for example, at the town of Arusha, small-scale farm equipment is designed and developed and then manufactured in workshops in the local centre. Ploughs and carts, harrows and seed planters, all designed to be pulled by oxen, are manufactured and repaired. These are more advanced than traditional tools, but less complicated than imported mechanical equipment. They are meant for use on the average farm in the country and suited to the needs and capabilities of the farmers. This sort of production is known as 'appropriate', 'intermediate' or sometimes 'village' technology. The big advantage is that the products are closely related to the needs of the users and are at a price they can afford. The workshops also provide many jobs. Many people claim that even in the industrialised countries massive, large-scale manufacturing has gone too far, and there should be a return to smaller factory units and less mass production.

1. What do the words 'appropriate' and 'intermediate' mean? Why are they good words to use to describe the sort of manufacturing found in many less industrialised countries?
2. List some of the advantages of 'intermediate' technology for people in the area concerned.
3. Describe any products from workshops or people in your town or area that could be described as 'intermediate technology'.
4. In the late 1970s the phrase 'small is beautiful' was sometimes used in the argument against modern giant companies and factories. What do you think the slogan means? What are some of the advantages of large size and large scale manufacturing?

Recycling metal and glass objects is a busy cottage industry in many countries. This photograph was taken on a street in Nairobi

Below: An example of 'appropriate' technology from Nigeria. These young men have made a tyre-repairing machine from an old lorry engine

13 Human skill

Making and using microcomputers

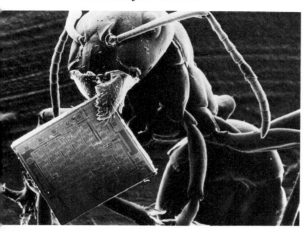

A micro-circuit so small that it can be held in the jaws of an ant

One of the most spectacular examples of human skill has been the development of the silicon chip. These tiny pieces of silicon are only a few millimetres square, but the integrated circuits on them are the basis of microcomputers and many different sorts of microelectronic products. They were first used in defence equipment and space vehicles in the early 1960s, but now their use is widespread. Equipment in offices and shops can store and recall information very quickly about sales, stocks, accounts, salaries and so on. Digital watches and pocket calculators are now commonplace and soon so will be the Prestel and Ceefax systems by which information can be received on television sets. In manufacturing industry the design, quality control and actual production with the help of 'robots' is now possible with the use of microcomputors. Few areas of life remain unaffected by developments in microelectronics.

In making silicon chips every process must be precisely controlled and carried out in conditions of absolute cleanliness. They are inexpensive because they are made in quantity. One wafer alone may produce enough chips to build three hundred microcomputers. The chips may be cheap, but the equipment required to make them is very expensive. Money for equipment and very high skill is as important in this manufacturing industry as money for raw materials. The electronics industry began in a big way in the Santa Clara Valley in California. Companies there had been making valves and transistors for years. Then new companies arose using this experience to make integrated circuits on silicon chips. In fact the valley became known

These women are operating machines which coat silicon wafers with a thin film of glass

Hundreds of micro-circuits are made from this single wafer of silicon

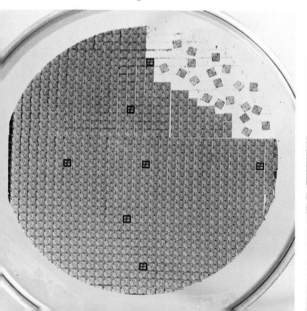

This machine 'scans' the drawing of an integrated circuit and feeds it into a computer. The image is then reduced to fit onto tiny silicon chips

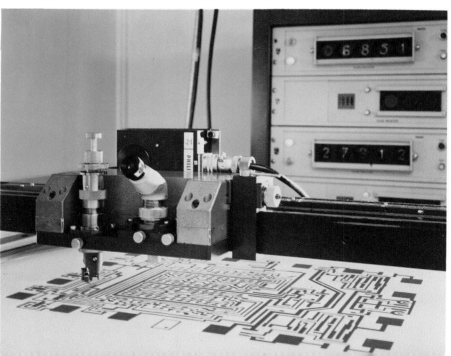

Left: Towns and cities in Ireland where foreign-owned microelectronics factories have been built

American owned companies in Ireland
Amdahl
Analog Devices
Centronics
Compugraphic
Computer Automation
Data Terminal Systems
Measurex
Memorex
Varian
Verbation
Wang Laboratories
Data 100
Data Products
Mostek
Prime Computers
Also
Nixdorf (Germany)
L.M. Ericsson (Sweden)
Nippon Electric (Japan)
Philips (Netherlands)

Mini Metro bodies travel down the robot assembly-line

One early product of the silicon chip revolution is the Prestel Information service

as Silicon Valley! Most modern industrial nations either have or are developing their microelectronics industry since they don't wish to rely completely on others for this essential need.

Many American, Canadian and Japanese firms want to locate microelectronics firms in Europe where labour is skilled, reliable and looking for new sorts of work. The Republic of Ireland has never been a major industrial country in the past, but it has plenty of labour, suitable sites for factories and a desire to support and benefit from this new science-based industry. A large number of foreign firms have been persuaded to open factories in towns and cities such as Galway, Limerick, and the Shannon airport industrial estate. They have done this by offering grants for factory building, training the workforce, research and development and by giving various sorts of tax relief. There are now over one hundred such firms. In the early 1980s, for example, Mostek Corporation of Dallas set up a plant in Dublin at a cost of over £40 million to assemble memory and microprocessor circuits. By 1983 a silicon chip making factory will be opened and towards the end of the 1980s over 1 200 people should be employed at the works. Ireland may not have been important for manufacturing in the past, but it clearly has a lot to offer the modern microelectronics industry.

1 Many more locations are suited to a micro-electronics factory than to a steelworks. Why is this?
2 Describe the scene inside the silicon chip factory. What do you think you would like and dislike about working in these sorts of conditions?
3 Make lists of five pieces of equipment used in the home, and five used in various sorts of transport, that depend in some way on electronics.
4 Some people will be employed in designing, making and controlling microelectronics equipment, but others will lose their jobs with the introduction of microcomputers. Describe some of the jobs where microprocessors may eventually take the place of workers.

14 Places as resources

Islands in the sun

Breakfast in style in the Caribbean (*above*) and then off to the pool and the beach (*right*), is the image that many tourists have of these islands. They don't see the sort of houses that many of the hotel workers live in (*below*)

Millions of people are fortunate enough to have the money and time away from work to travel as tourists to other parts of the world. They are attracted by many different things, but the scenery and climate are undoubtedly amongst the more important of these.

In the Caribbean Sea, between the USA and the South American mainland, lie a large number of islands that have experienced a big growth in tourism in recent decades. Most of them are within the Tropics, and the expectation of hot sunny weather is one of the reasons why many tourists go there – especially during the American and European cold winter months. It is not widely realised, though, that it can be very rainy, particularly in the mountainous volcanic islands. The islands are also sometimes crossed by hurricanes that bring violent weather. Even so, people continue to travel to the Caribbean in search of the sun.

Many of the islands are not volcanic, but consist of low-lying sedimentary rocks. They often have beautiful sandy, palm-fringed beaches leading to clear blue seas. Offshore are many coral reefs and islands, so swimming and sailing and fishing are popular. To provide accomodation for the tourists many hotels have been built, while on the larger islands modern sea and air ports allow large numbers of tourists to arrive and depart. Many people obtain employment through the tourist industry – such as in hotels and driving taxis. The tourists also spend their money on food, entertainment and souvenirs. Because of all this money that is spent and the jobs that are provided it looks as though the environment is a valuable resource that brings nothing but good. But some people are not so sure.

In the first place most of the hotels are foreign owned and the top jobs are often taken by foreigners. Local people may be employed as waiters, cooks, cleaners and so on. Much of the money is spent within the hotels. Most of the food eaten is imported and not bought from local farmers. Much of the income from tourism, therefore, goes to foreign tour operators, travel agents, airlines and hotel companies.

Opportunities for jobs are not so straightforward, either. The tourist trade is seasonal, and many of the people employed in the tourist industry are unemployed during the quiet season. As everywhere else in the world, mass tourism can spoil the attractions of the environment. Some of the best beaches are owned by the hotels and cannot be used by the local islanders, while on some islands a lot of land is owned by wealthy people in other countries. There is no doubt that the environment is a valuable resource – but a lot of the money earned from it does not benefit the ordinary islander.

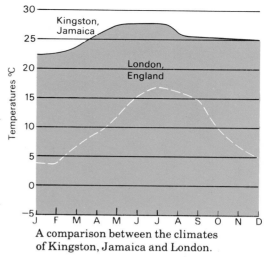

A comparison between the climates of Kingston, Jamaica and London.

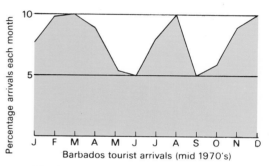

Barbados tourist arrivals (mid 1970's)

The frequency of tourist arrivals in Barbados throughout the year

1. Look at the temperature graphs for Kingston and London, and the seasonal pattern of tourist arrivals for Barbados. In which months did most tourists arrive? Why do the peaks appear at these times of the year?
2. From the diagrams suggest why a) so many tourists in the Bahamas are from the USA, b) so many French tourists visit Martinique while far more Canadians and British visit the neighbouring island of Barbados.
3. Describe the scene on the left. What are the attractions of the local environment? What have people added to these to increase the attraction of the place for tourists?
4. How could the local islanders share more in the benefits of tourism?

The proportion of foreign visitors to different West Indian islands

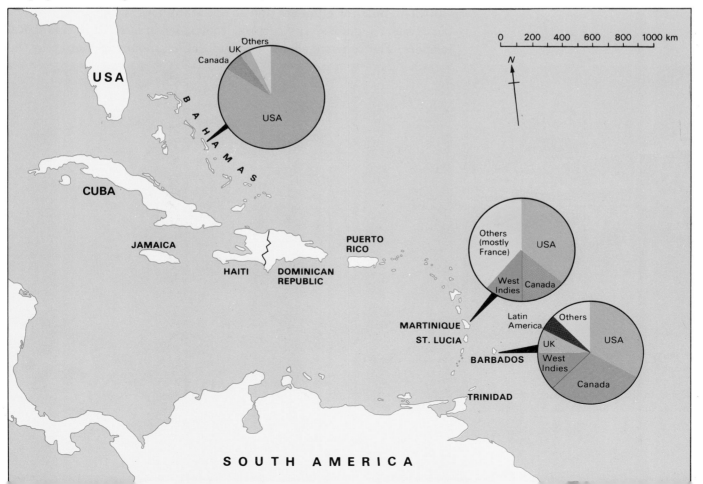

14 Places as resources

Holidays in the Alps

Summer is a good time to go walking in the mountains in Switzerland

Many people choose to take their holidays in mountain areas. Some are attracted by the spectacular and beautiful scenery with its mixture of snow-covered peaks, sharp-edged ridges, glaciers, valleys and lakes. The landscape changes dramatically with the seasons. In winter everything is covered with deep snow. In the summer when the snow has melted everywhere except on the highest peaks, there is the enjoyable contrast between the high mountain grassland on the flatter land beneath the mountain peaks, the wooded hillsides and the valley floor. Here there is the scatter of wooden farmhouses, cultivated fields and small villages. Each season attracts its visitors for different reasons. There are many such tourist areas in the Alps in Switzerland, Austria and North Italy, and examples of winter and summer landscapes can be seen on these two pages.

Not all people visit the mountains just for the attractive scenery. Some enjoy more vigorous activities such as mountain walking, climbing or skiing. Once people started to visit the mountains the local inhabitants began to build accommodation for them in chalets and hotels. Some towns such as St Moritz are full of hotels and restaurants that cater for the many tourists. All sorts of other amenities have been provided to attract visitors – ski-lifts, mountain railways, huts for climbers and so on. Many local people work full or part-time as guides, instructors, coach drivers, waiters and hotel keepers. The tourists spend a considerable amount of money in the

A party of British tourists pose by their coach in the Alps

Right: In the winter thousands of holidaymakers come to the Alps to ski

St Moritz is one of the most famous (and most expensive!) ski resorts in the world

area. Although some activities such as skiing or hill walking are seasonal, the variety of attractions means that many of the tourist areas have visitors thoughout the year.

It is worth recalling that the Alps have not always been tourist areas. Mountains have not always been regarded as pleasant or enjoyable places. We know from books and letters that until about two hundred years ago many people found mountains dangerous, ugly and frightening. It is only more recently that these environments have appeared beautiful, inspiring and inviting and the clear, fresh air thought to be very healthy. It is fascinating to see how the attitude of people to mountain environments has changed so much over time.

It is also only recently that there has been a good network of road and railway within the Alps to allow people to visit them. Nowadays it is possible to fly by jet into many of the main cities and go by road or rail into most mountain areas quickly and in comfort. So the mountains have become accessible to large numbers of tourists. The mountain environments of the Alps are important resources for the local people and the countries within whose borders they lie.

The Alpine lakes are another major attraction for tourists. This is Lake Lugano

1. Copy the diagram of the Alpine mountain and valley landscape and add these labels in the appropriate places: mountain peaks; summer snowline; Alpine meadows; coniferous forest; U-shaped valley; farm and village.
2. Either describe a holiday you have taken in a mountain area, or describe what sort of activity would most appeal to you if you had the chance to take one.
3. List all the ways in which tourists to a mountain area might spend their money or list all the sorts of employment that might be available as the result of tourism.
4. Mountain scenery is not a non-renewable resource like coal or oil, so it will never be 'used up'. But there are some dangers that might stop it being a valuable resource. Suggest a few of the dangers to tourism in mountain areas.

An Alpine mountain and valley landscape

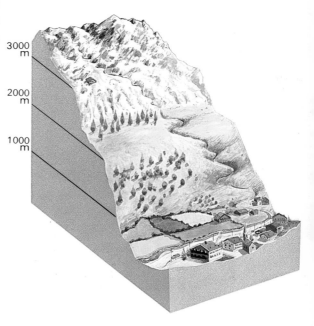

14 Places as resources

East African game parks

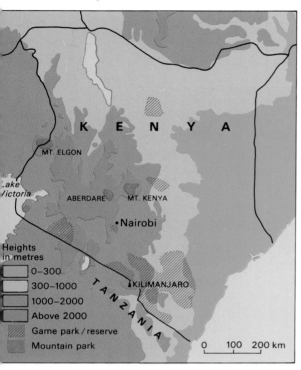

Kenya

Below: Strange and beautiful animals such as these are what the visitors come to Kenya to see

A great deal of tourism in East Africa is based on the marvellous variety of wildlife. In Kenya about seven per cent of the country is used as wildlife parks and reserves. These vary in size, and are scattered throughout the country. The largest are the savanna parks of Tsavo and Masailand in the south. Many sorts of animals and birds are found in these environments of grassland and scattered trees, while the spectacular scenery provided by Mount Kilimanjaro and the steep slopes of the Rift Valley add to the attractions. The drier northern reserves and the mountain parks are places where different sorts of wildlife may be found.

Kenya became an independent country in 1964. Between then and 1980 earnings from tourism increased sevenfold, and tourism is now second only to coffee as a source of foreign money. People travel there from all over the world. The largest numbers are from North America and Europe, but the Japanese and Australians are the biggest spenders as individuals. Kenya has benefited from the world-wide growth of the package tour holiday, and about half the tourist income is from this type of visitor. The airlines and hotels welcome package tourists, but they tend to want to see the main sites as quickly and cheaply as possible and so put a lot of pressure on the accessible parks at peak times of the year. The government tries to encourage individual tourists who tend to stay longer although fewer in number, to travel more widely and spend more money per person.

The government realises the importance of tourism and has carefully planned its growth. As well as increased numbers of hotels in the main towns there are now different sorts of game lodges in the reserves themselves. These have usually been designed to merge into the landscape or based on traditional building designs seen in the Masai villages. The aim is to encourage game viewing – big-game hunting is now prohibited in Kenya.

Because land is badly needed for farming and ranching it is not likely that more reserves and parks will be created. The aim is to use the existing parks more efficiently, to conserve the wildlife and the environment and to prevent the damaging effects of too much tourism. Over 1 000 kilometres of all-weather track are being built to encourage visitors to move about the parks. This will help reduce 'horizon pollution' where it becomes impossible to see any animals without a cluster of motor vehicles around them! Tourism has brought valuable employment and money to the country. The government is keen to conserve the wildlife and environment so that these resources continue to give pleasure to people from all over the world and benefit to the people of Kenya.

Nowadays people hunt for game with their cameras. These tourist vans are standing on the plains beneath Mt Kilimanjaro in Tanzania

1 a) What important line of latitude passes through Kenya? b) What sort of mountain do you think Kilimanjaro is, from its shape? c) How do you explain that mountains in these latitudes are often capped with snow?
2 In what ways might 'package' tourists put more pressure on the wildlife and game reserves than visitors travelling as individuals or in very small groups?
3 Describe one of the scenes and say why you would like to be there.
4 Resources are often said to either renewable (there is always more where it comes from) or non-renewable (once the resource is used then it is gone forever). Is the wildlife and landscape of an East African reserve a renewable or a non-renewable resource – or a bit of both?

The Ngorogoro Crater in Tanzania is famous as a home for wild life. The pink flamingoes are a big attraction

14 Places as resources

Places with a past

This advertisement for a Mediterranean cruise puts emphasis on the historic places that the liner will be visiting

THE CRUISE THAT'S TAKEN 5,000 YEARS TO PREPARE.

The 'Cradle of Civilisation' cruise on Queen Elizabeth 2 next year will be an historic event in more ways than one. Besides enjoying a luxurious holiday on the greatest liner afloat, you will visit some of the most ancient and enthralling places in the world.

In 3,000 BC the Pyramids were finally completed. An achievement made even more impressive when you remember the only tools available were the slaves.

Above Lisbon stands the Tower of Bélem – a fine example of the flamboyant Gothic style. Here you can stand at the point where famous navigators like Vasco da Gama set sail on voyages of discovery.

1515 A.D. TOWER OF BÉLEM – LISBON

The intriguing city of Istanbul was the crossing place to Asia Minor for Greeks, Romans, Crusaders and Turks.

The many different influences on the architecture have created a fascinating assembly of arches, turrets and towers.

Many of our architectural principles were provided by the Greeks. And the ultimate showpiece of their craftsmanship,

600 B.C. ACROPOLIS – ATHENS

the Acropolis, still stands majestic above the ancient city of Athens.

The city of Jerusalem witnessed many events of the Old Testament.

And today you can explore the narrow streets that have remained almost unchanged since the days of the early Christians.

691 A.D. DOME OF THE ROCK – JERUSALEM

2,000 years ago in Pompeii, time stopped. The volcanic ash has been carefully removed and the city now stands perfectly preserved.

The people's positions, frozen as their world ended, have been recreated for the future in plaster of Paris.

Final preparations for this cruise were completed in 1967 when the QE2

was launched by Her Royal Highness Queen Elizabeth II.

The ship has six sun decks, four pools, eight cocktail bars, three restaurants, a cinema, a ballroom, nightclub and casino – the list is almost endless.

79 A.D. POMPEII

In fact the QE2 has everything you could desire to make the 'Cradle of Civilisation' the most civilised cruise in history.

The cruise starts on April 2nd '76 with a flight from London to Lisbon to join the QE2.

1967 A.D. QE2 – SOUTHAMPTON

Then to Palma, Alexandria, Istanbul, Athens, Odessa, Haifa, Naples, Cannes and the last port of call, Barcelona, where a jet will be waiting to fly you home.

After 5,000 years in preparation, three months to book isn't very long.

For full details phone Cunard Leisure (01) 491 3930 or (0703) 29933, cut out the coupon or contact your travel agent.

For full details of the 'Cradle of Civilisation' cruise April 2-24 1976, send this coupon to Cunard Leisure, c/o Pembroke House, Campshourne Rd., London N8 7PT.

(BLOCK CAPITALS)

Name

Address

Tel:
G ICC

CUNARD QE2

640 – 1453 A.D. – ISTANBUL

It is perhaps surprising that even when people have had to struggle hard to survive in a harsh environment they have often created beautiful or spectacular buildings such as monuments, temples and cathedrals. Sometimes the building was done at great human sacrifice for the glory of a ruthless ruler or a wealthy or powerful person, but often it was for the benefit of the whole community.

Much of the past has been destroyed or just disappeared with time, but here and there evidence of the past exists for us to see and explore and wonder about. We are attracted to some old buildings by their beauty and that of their contents, or by the knowledge that people (often famous) lived, worked and relaxed in them. But some sites have not survived very well and are not very spectacular to look at. Their attraction comes out of realising that men and women built and lived in them thousands of years ago and that some of the evidence has survived over many generations.

Buildings wear away with time under the normal and natural processes of weathering by the wind, rain, frost and heat. In recent years the stone and brickwork of some old buildings has been attacked with disastrous results by chemicals in the atmosphere produced by industry and traffic. While some sites and historic buildings are in the countryside many are in modern, bustling cities. Although they are too important ever to be destroyed for new development, they may be swamped by new buildings such as high-rise offices and flats or encircled by urban motorways carrying vast amounts of traffic. The settings of the buildings may be spoilt and the fabric damaged by all these activities. Finally the very attraction of the historic buildings may result in so many visitors that it stops being a pleasure to see it – quite apart from the clutter of ice cream, food stalls, and souvenir stands and so on that destroy the sense of being in the past or seeing the place as it once was.

The Tower of London is a popular place for tourists (and school parties!). This notice turns an ordinary patch of grass into an historic place by stimulating peoples' imaginations

This past heritage covers many sorts of sites and buildings all over the world and from all periods of time. Not all are equally spectacular, important, interesting or acceptable, but in a world of rapid change it is important to have some reminders of our past. Tourism is important because it provides employment and money for the people and country where the resources are located. Even if this were not so, as with beautiful scenery and wildlife it would be criminal to destroy evidence of our past.

1. Choose one of the historic buildings shown here. Describe the building and its setting, and say what it is that makes it an attraction to so many tourists. If you think none of them is interesting say what you would do with them or use them for.
2. Name any historic sites that you have visited or know about that are popular because **a)** someone famous lived there **b)** an important event such as a battle or treaty or conference took place there, **c)** it is an impressive defensive building, **d)** it is an impressive religious building, **e)** there are few traces of buildings but it helps us imagine what the past was like thousands of years ago.
3. List some of the dangers to historic sites and buildings, and say what could be done to protect them from the dangers you mention.

15 Places, resources and people

Signs of strain

A South American woman working in her 'kitchen'. People like us would find this kind of poverty hard to understand

Below right: These starving mothers and children are victims of both the drought and the war that affected Uganda after the defeat of Idi Amin

Although standards of living are rapidly improving in parts of Asia, overcrowding remains a problem

Ever since men and women appeared on earth they have affected the environment in which they lived and used its resources as best they could with the knowledge and skill they had. From these earliest days many people have found it difficult to live well, avoid disease and hunger and escape natural disaster or warfare. The strains and dangers that can be seen today are not new. What is worrying is the extent of disasters and damage and the speed with which they seem to be getting worse.

In spite of all the developments that have taken place, and all the modern technology available, the Food and Agriculture Organisation reported in the middle of 1979 that twenty countries were affected by abnormal food shortages for some reason or other. In the early 1980s it was thought that more than one thousand million people suffered from having too little food and perhaps as many as 400 million lived on the brink of starvation every day of their lives. It is estimated that between ten and twenty million people die of starvation or its effects every year – but no one knows for sure.

It is difficult for people living in Britain to understand what this sort of malnutrition and fear of starvation is like. Most of them get used to all the reports and pictures of disaster and get on with their own lives. Malnutrition and starvation are unevenly spread around the world. A fairly accurate measure of malnutrition is the number of deaths of pre-school children. In some parts of Africa, Asia and Latin

David Attenborough has joined in the plea for conservation of the world's tropical rain forests. He points out that these forests are the earth's 'lungs' and without them severe damage to the world's climate patterns would take place – and yet they are being destroyed at the rate of 20 hectares a minute!

The Times 6 June 1980

Battle for oil may end in war, says minister

The scramble for oil could end in a world war, Mr. David Howell, the Energy Secretary, said yesterday. The developed world's independence on oil must be drastically reduced.

In one of the bluntest statements on the dangers inherent in the present energy crisis, Mr. Howell said that the developed world's dependence on oil had to be drastically reduced. It was especially important for the West to cut back on the oil it took from the Gulf states, any one of which could explode at any minute.

Mr. Howell's remarks come only days before members of the Organization of Petroleum Exporting Countries (Opec) meet in Algiers to discuss future pricing policy. He gave a warning that it would take the Western economies, built on cheap oil, many years to adjust to the changed world of tight supplies and rising prices.

This dependence on oil cast a triple shadow across hopes for world stability and peace. It put pressure on supplies and caused damage to Western economies. Enormous pressures were placed on British society by the need to absorb large price increases.

Secondly, oil dependence murdered, crucified the development plans of the developing countries. Tight supplies hampered their development.

Thirdly, it set nation against nation in a struggle for too little oil. East and West could jostle each other with increasing roughness in a desperate sort of international musical chairs. This could eventually lead to war.

America these can be up to forty times higher than in the USA or Western Europe. Malnutrition does not always lead to starvation and death. In most cases it leads to never-ending weakness, illness and disease that makes life difficult and unpleasant. On top of all this many of the people who do not have enough food also have inadequate houses to live in. In the worst cases they have nothing at all, and families have to sleep on the streets or in shelters of cardboard, wood or other waste materials.

It would be quite wrong to think that all the signs of strain are in the parts of the world where the poorest people live. Many environments nearly empty of people are being destroyed for their resources. As the caption above says, this will have disastrous effects unless checked soon. Other resources such as oil are so vital to the wealthy industrialised nations that they are beginning to fear disaster will happen if supplies are cut off. In the meantime man-made cities are getting bigger and more congested, the environments in the wealthy areas becoming more and more polluted and diseases caused by overeating or bad diets becoming major causes of early death. The signs of strain are world-wide.

1. Look at the picture of the woman in the 'room' washing clothes. Compare it with the room in your house where the washing is done.
2. List reasons for conserving the world's forests.
3. What are some of the 'signs of strain' in Britain that you have seen or heard about on television or in newspapers?
4. What do you think the more fortunate countries should do to help those with many poor and hungry people?

15 Places, resources and people

Unequal shares

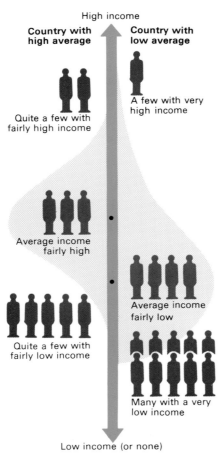

A comparison of incomes in two parts of the world

These two posters, found side by side in an English street, speak for themselves

There are many ways of measuring the wealth or well-being of people. It is true that material wealth alone does not necessarily make people happy. But it is hard for those who are hungry or starving, ill or diseased, unemployed or slaving long hours merely to survive to feel satisfied or have much hope for their future.

One obvious measure is the amount of food there is to eat. The two advertisements accidentally placed side by side on the hoarding emphasise that there are enormous differences from place to place. The girl on the right represents the majority of people in Western Europe and North America, white South Africa and Australasia. The one on the left represents the millions of less fortunate in Africa, Asia and Latin America. It is very easy to get used to such images and forget the real people they represent. The poster on the left has been produced by some English people in a voluntary organisation trying to raise money to help those in need.

The map tries to make a similar point – that there is an unequal distribution of foods, in the form of calories, available to be eaten each day. Both the advertisements and the map have to be read with caution, though. They only show what the average conditions are like, and give no idea of the calorie intake of the wealthiest and poorest people in each country. But if we bear this in mind the map does allow a genuine comparison to be made.

Another measure of the wealth of people is the sort of houses they live in. Some have enough money to live in newly built and lavishly equipped flats like these in Paris. Others have to live in the miserable squalor of slums like these in Bombay. But it is too simple to say that some countries are wealthier than others. There are areas of very poor housing in Paris, and the slum dwellers in Bombay have the background of the high-rise developments to remind them of the considerable wealth owned by some Indians. There are contrasts of wealth and poverty within countries as well as between them.

It is sometimes said that on the world scale there is no shortage of food or housing, resources nor energy. In some places there is more than enough and a great deal of waste is accepted. At the very same time other people in other places go hungry and live in extreme poverty without any hope of improvement. A big question is why these differences exist – and many different answers are given. Another question is what should the more fortunate individuals and countries do about it, if anything. Should they give aid or offer trade or get on with their own problems? The last answer seems shortsighted. As an international report made in 1980 stated 'the industrialised and developing worlds are totally interdependent.'. In the end, no country can remain unaffected by the rest of the world.

Although overcrowding is a problem in the West, there is often enough money for solutions to be found. The picture above shows modern high-rise flats in Paris. India (*below*) has its luxury high-rise flats too (seen in the background) but the people who build them live in the makeshift huts in the foreground

1 From the map name two continents without any country having less than the desirable calorie intake, and one without any country having above the desirable intake.
2 On this map the sizes of the shapes represent the populations of the countries, not their areas. Name three parts of the world where the average calorie intake is low.
3 Imagine you were one of the people in the advertisements looking at the other person. Say which you are. What do you think you would be feeling and thinking?
4 Suggest a few reasons why there are differences of wealth and living standards around the world.
5 Suggest some ways in which 'the industrialised and developing worlds are totally interdependent'. (Look back through this book for some clues)

Daily calorie consumption per head

Over 2900	Above desirable amount
2500–2900	
2100–2500	Below desirable amount
under 2100	

Key
1 Canada
2 USA
3 Mexico
4 Venezuela
5 Brazil
6 Argentine
7 United Kingdom
8 Sweden
9 West Germany
10 Italy
11 Nigeria
12 Egypt
13 South Africa
14 USSR
15 China
16 India
17 Bangladesh
18 Taiwan
19 Japan
20 Philippines
21 Indonesia
22 Australia
23 New Zealand

Left: A 'topological' world map of the daily consumption of calories per head

15 Places, resources and people

Too many people?

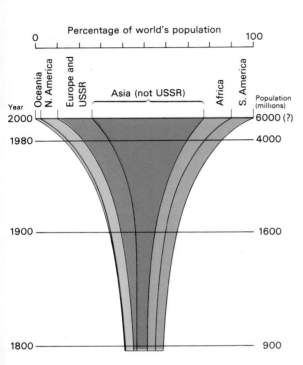

The distribution and growth of world population between 1800–2000

Rush hour on a Bombay train. This is rather like rush hour on our trains – only worse

The Philippines has one of the most rapidly-expanding populations in the world, yet it would be unfair and untrue to say that most of its people are unhappy or unhealthy

One explanation for the millions of poor in the world is that there are just too many people. Those who say this usually mean that a country hasn't enough resources to feed and employ its people. They also point out that it is usually in these poorer countries that population is growing very rapidly. If they are right the only solution would seem to be to halt the growth of population – by persuasion or force.

Others think this explanation is too simple. There are some countries with a fairly high standard of living where many people live in a small area – where the density of population is very high – and natural resources are limited. These people manage by trading with other countries. They sell raw materials, energy, manufactured goods or services and with the money this earns they buy the things they need. Britain and Japan are two examples, while Hong Kong is perhaps an even better illustration. The belief is that the poorer countries could do the same if only the richer ones would help them get started.

The map and diagram show that the population of the world is unevenly distributed and growing very rapidly. A number of things affect the population size of a country. The most important of these are the numbers of births and deaths, and the numbers of migrants who enter and leave the country. Together, these result in a gain or loss of population.

Improvements in food supply, medical care and housing usually result in fewer people dying – and an increase in population! These developments are happening, fortunately, in many poorer areas. In most western countries adults choose to have small families and have the knowledge and means to practise birth control. Some don't do so for religious or personal reasons, and choose to have large families. In the poorer countries many adults know little about family planning, or if they do they cannot afford to practise it. They often choose to have large families anyway, in spite of their poverty. An extra pair of hands can help support the family, look after parents in old age and possibly satisfy religious beliefs about family life. Even so, in most developing countries the governments encourage and support family planning. In spite of great difficulties there are signs that birth rates are beginning to fall in some of these.

The fall in birth rates alone, however, will not solve the population problem. As both birth and death rates fall, the age structure of the population will change and there will be more elderly people and young children to be supported. Population control has to go hand in hand with other sorts of development.

1. What percentage of the world's population lives in **a)** Europe and the USSR, **b)** Asia?
2. Draw two columns to represent the populations of two countries – one of 50 millions and one of 100 millions. If they both increase at the same rate – for example, they double over ten years – show how many people will then live in both countries. Add the note: the rate of change can be the same but the extra numbers can be very different!
3. What are some of the problems of living in a high-density country even if the standard of living is quite high? What are some of the dangers of relying on trade with other countries for a high standard of living?
4. Imagine you are an adult either in a developed country such as Britain or in a less developed one such as India. For the one you choose say what would affect the size of family that you would like to have.

Refugees from Vietnam. Their boat is seen here being towed into harbour. Large numbers of these 'boat-people' have made Hong Kong more crowded than ever

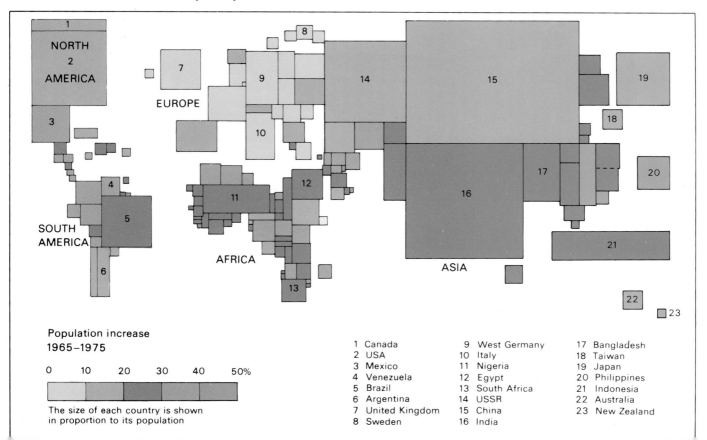

World population distribution and change, 1965–75

Population increase 1965–1975

0 10 20 30 40 50%

The size of each country is shown in proportion to its population

1 Canada
2 USA
3 Mexico
4 Venezuela
5 Brazil
6 Argentina
7 United Kingdom
8 Sweden
9 West Germany
10 Italy
11 Nigeria
12 Egypt
13 South Africa
14 USSR
15 China
16 India
17 Bangladesh
18 Taiwan
19 Japan
20 Philippines
21 Indonesia
22 Australia
23 New Zealand

15 Places, resources and people

Looking ahead

In the Indian State of Uttar Pradesh, a class of girls enjoy a reading lesson. In many countries far more women than men are illiterate. This is because the education of women has been thought of less importance than that of men

Lectures like this, teaching Africans to recognise the signs of smallpox have helped to wipe out this disease

There are worrying signs that environments, resources and people are under increasing strain. Many environments that took millions of years to evolve are being destroyed for ever as they are 'opened up' in the search for, and development of, resources. Whether this is being done because of urgent need or for greed makes little difference to the effects. The trouble is that apart from the destruction, damage may be being done to the whole web of life. The consequences of interfering with the environment on such a huge scale are just not known.

For decades the industrial countries of the world have been able to get cheap energy, and it has been used wastefully. It is now realised that energy and mineral resources will not last forever. Indeed some are likely to be exhausted over the next few decades at the present rate of use. Great efforts are beginning to be made to use these scarce resources more carefully, and to seek alternative forms of energy. If this fails the next two decades could be dangerously short of vital resources.

The world's population continues to grow at an alarming rate, although there are signs of a slight slowing down. If population growth is not slowed down, especially in the poorer countries, it will be difficult to produce enough food to feed everyone, even with improved methods and increased production. The gap between the rich and poor within countries and between countries doesn't seem to be getting smaller. In the early 1980s there were many sad examples

of these differences. The fears and resentments they produce in rich and poor alike, can lead to violent protests and even wars.

But there are also some encouraging developments. Efforts to control population growth are beginning to have some effect, while many countries once controlled by others have now gained their independence and control their own affairs. There continue to be great advances in food production, medicine and technology that make it possible for everyone to have enough to eat, be free from many diseases and have a share of resources to make life better than a never-ending drudgery.

Most important of all, though, is the attitude of people. It is increasingly realised that in many ways every part of the world is affected by what happens elsewhere. It is becoming more and more difficult for one country to use the resources and the people of another and ignore it at the same time. The greatest need and the only long-term hope is for us to care about the environment, its resources and about the well being of other people – and to do something about it.

1. Look back through this book and give four examples of the way in which the natural environment has been drastically affected by people in recent years. Describe the changes and some of the damage and dangers resulting from them.
2. Name two sorts of minerals that are thought to be near exhaustion. What are countries that need them doing about this expected loss of supplies?
3. Draw diagrams to show the following facts: **a)** Britain's aid to developing countries in 1980 = 0.4 per cent of its wealth produced during the year. **b)** aid for overseas programmes from Britain in 1978 = £26 million from voluntary agencies; £726 millions from taxes.
4. The four young people below have said what they most look forward to in the year 2000. Write your own short statement of what you would look forward to most for yourself.

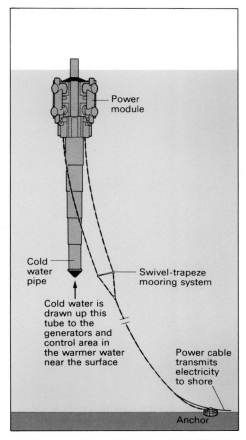

An ocean thermal energy plant. This is a means of creating energy from the difference in temperature between the upper and lower levels of the sea.

The houses will be made of stone. They will cost $250 and will last a long time. The roads will be made of tar. There will not be any beggars and poor people. Everyone will have jobs and food.
Asanatu Koroma (13) Sierra Leone.

I don't want to be rich. People just envy you. Why should you have more than you need. Once you've got a home, family, friends and a job you should be satisfied.
Desmond Thomas (14) Jamaica.

The endless rows of trees will be gone and there will be barbed wire instead. Really, I don't think there'll be anything after the Year 2000. I just think it will stop then.
Bjorn Kellman (15) Sweden.

I don't ever want to go broke. The husband I have would have to be in pretty good shape moneywise because of the lifestyle I've grown up with. I want the same level of living because it's hard to drop down.
Tracy Cernan (15) U.S.A.

Children from different parts of the world describe 'My World in the Year 2000'

Acknowledgements

The publisher would like to thank the following for permission to reproduce photographs:

Aerofilms, p. 75; Aramco World, pp. 24, 27, 28, 29; Aspect, pp. 49, 65, 68, 82, 83, 115, 117, 120; Associated Press, p. 36; Atlas Copco GB Ltd., pp. 13, 36; Barnaby's, pp. 39, 79, 114; BBC Pictorial Publicity, p. 121; Rex Beddis, p. 66; Black Star New York, p. 54; J. F. E. Bloss, p. 69; BP, pp. 96, 100, 101; British Leyland, p. 111; Len Brown, p. 109; Romano Cagnoni, pp. 88, 98, 105; Camerapix, p. 86; Camera Press, pp. 39, 40, 44, 79, 81, 94; J. Allan Cash, p. 108; H. J. Harrison Church, p. 70; COI, p. 110; Bruce Coleman, pp. 19, 68, 86, 91, 93, 116; Crown Copyright, p. 104; Cunard Line Ltd., p. 118; Daily Telegraph Colour Library, pp. 37, 47, 66; De Beers, p. 31; Mason Dooley, p. 49; David Drewry, p. 46; R. J. Dunn, p. 122; Mark Edwards, p. 33; Tor Eigeland, p. 25; Elisabeth Photo Library, p. 73; ESA, p. 10; Robert Estall, p. 112; Mary Evans Picture Library, p. 92; Everetts Ltd., p. 22; Ewbank and Partners Ltd., p. 65; Fiat Auto S.p.a., pp. 106, 107; Fay Godwin, p. 96; Goodyear Aerospace, p. 63; GPO, p. 111; Richard and Sally Greenhill, p. 72; Susan Griggs, pp. 21, 26, 27, 42, 60, 75, 76; The Guardian, pp. 95, 119; Tom Hanley, pp. 39, 76, 120, 124; Robert Harding, pp. 17, 29, 51; John Hillelson, pp. 14, 16, 20, 34, 48, 53, 55, 72, 82, 101, 126; Michael Holford, p. 56; Hong Kong Government, p. 125; Michael Howarth, pp. 57, 58; Alan Hutchison, pp. 16, 21, 29, 30, 35, 56, 64, 69, 82; International Wool Secretariat, p. 81; Raymond Irons, p. 88; Tessa Jackson, p. 56; Marion Kaplan/Camera Press, p. 126; Kennecott Minerals Company, p. 102; Hugo von Lawick, p. 15; Library of Congress, p. 52; London School of Hygiene and Tropical Medicine, p. 83; The Mansell Collection, p. 86; Cherry Mares, p. 123; Steve McCutcheon, p. 45; MEPLA/D. Shirreff, p. 28; Miggs Morris, pp. 43, 44, 45; Marion Morrison, p. 87; Tony Morrison, pp. 21, 58, 59, 89; Mullard Ltd., P. 110; Margaret Murray, pp. 32, 33, 37, 108, 109, 123, 124; Musée de l'Homme, p. 80; NASA, p. 67; Nebraska State Historical Society, p. 52; Thomas Nelson and Son Ltd., p. 60; New Zealand High Commission, pp. 84, 85; NHPA, p. 90; Novosti, p. 70; N. V. Philips, GloeilampenFabrieken, p. 110; The Observer, pp. 97, 104; The Observer/Camera Press, pp. 20, 94; P and O Landtours, p. 114; Pix Features, p. 36; Rapho, p. 11; Roan Consolidated Mines, p. 103; Bryan Sage, p. 42; Seaphot, p. 90; James Siers, p. 85; Smithsonian Institution, p. 52; Space Frontiers Ltd., pp. 8, 9, 12, 55; Spectrum, pp. 18, 38, 62, 66, 76, 78, 112; Frank Spooner, pp. 13, 62, 93; David Sugden, p. 46; The Sunday Times, p. 98; Swedish Tourist Office, p. 80; Swiss National Tourist Office, p. 114; Chris Sykes, p. 108; Syndication International, p. 121; Jeffrey Tabberner, pp. 63, 115; David Taylor, p. 57; The Times, p. 95; John Topham, p. 31; UAC, p. 23; J. P. Weber, p. 30; WHO, pp. 18, 64, 120; World Wildlife Fund, p. 101; George Young Photographers, p. 94; ZEFA. pp. 49, 50, 105.

The illustrations are by Gary Hincks/Institute of Geological Sciences, pp. 9, 10–11, 63; Pinpoint Graphics, pp. 8, 9, 14, 19, 38, 43, 72, 73, 77, 81, 87, 107, 115; Richard Smith, p. 41; Ann Usborne, p. 127.

Picture Research by Ann Usborne.